Geochemical Rate Models

An Introduction to Geochemical Kinetics

Rate models (kinetic models) are used to predict rates of geochemical processes in the near-surface environment. Based on the author's 30 years of teaching and research experience, this combination of reference and textbook provides a systematic, comprehensive description of rate models, developed from fundamental kinetic theory and presented using consistent terminology and notation.

Major topics include rate equations, reactor theory, transition state theory, surface reactivity, advective and diffusive transport, aggregation kinetics, nucleation kinetics, and solid–solid transformation rates. The theoretical basis and mathematical derivation of each model is presented in detail and illustrated with worked examples from real-world applications to geochemical problems. The book is also supported by online resources: self-study problems put students' new learning into practice; and spreadsheets provide the full data used in figures and examples, enabling students to manipulate the data for themselves.

This is an ideal overview for advanced undergraduate and graduate students, providing a solid understanding of geochemical kinetics. It will also provide researchers and professional geochemists with a valuable reference for solving technical and scientific problems.

J. Donald Rimstidt is a Professor Emeritus of Geochemistry at Virginia Polytechnic Institute and State University (Virginia Tech). During his 30-year career there he has taught several graduate courses, including Aqueous Geochemistry, Environmental Geochemistry, Geochemical Kinetics, and Geochemical Thermodynamics. The effective pedagogic methods from those courses, for building quantitative skills and instructing modeling techniques, underlie this book. Professor Rimstidt is the author or co-author of more than 100 papers, including his seminal article, "The kinetics of silica-water reactions" (with Hu Barnes, 1980), which forms the basis of the quasi-thermodynamic presentation of transition-state theory presented in this book.

"Don Rimstidt brings 30 years of teaching experience at a world class geosciences department to the writing of this book. With its beautifully clear exposition of the theoretical and practical aspects of geochemical kinetics, and use of real-world examples, this book is essential reading for students and professionals with interests in geochemical rate models."

David J. Vaughan, *Research Professor, University of Manchester, and President of the Mineralogical Society of America*

Geochemical Rate Models

An Introduction to Geochemical Kinetics

J. Donald Rimstidt

Virginia Polytechnic Institute and State University

CAMBRIDGE
UNIVERSITY PRESS

University Printing House, Cambridge CB2 8BS, United Kingdom

Published in the United States of America by Cambridge University Press, New York

Cambridge University Press is part of the University of Cambridge.

It furthers the University's mission by disseminating knowledge in the pursuit of education, learning and research at the highest international levels of excellence.

www.cambridge.org
Information on this title: www.cambridge.org/9781107029972

© J. Donald Rimstidt 2014

This publication is in copyright. Subject to statutory exception and to the provisions of relevant collective licensing agreements, no reproduction of any part may take place without the written permission of Cambridge University Press.

First published 2014

Printed and bound in the United Kingdom by TJ International Ltd., Padstow, Cornwall

A catalogue record for this publication is available from the British Library

Library of Congress Cataloguing in Publication data
Rimstidt, J. Donald, 1947– author.
Geochemical rate models : an introduction to geochemical kinetics /
J. Donald Rimstidt, Virginia Polytechnic Institute and State University.
   pages cm
Includes bibliographical references and index.
ISBN 978-1-107-02997-2 (hardback)
1. Geochemical modeling. 2. Chemical kinetics. 3. Geochemistry–Mathematical models. I. Title.
QE515.5.G43R56 2014
551.9–dc23   2013022101

ISBN 978-1-107-02997-2 Hardback

Additional resources for this publication at www.cambridge.org/rimstidt

Cambridge University Press has no responsibility for the persistence or accuracy of URLs for external or third-party internet websites referred to in this publication, and does not guarantee that any content on such websites is, or will remain, accurate or appropriate.

For my students

Contents

Preface	*page* ix
1 Geochemical models	1
2 Modeling tools	9
3 Rate equations	36
4 Chemical reactors	56
5 Molecular kinetics	79
6 Surface kinetics	102
7 Diffusion and advection	128
8 Quasi-kinetics	156
9 Accretion and transformation kinetics	182
10 Pattern formation	205
References	210
Index	228

Preface

This book is intended for use as a textbook for an upper-level undergraduate or a first-year graduate course and as a handy reference for professionals who need to refresh their memories about a particular rate model. It focuses on rate models that are well understood, widely used, and pertinent to geochemical processes. Its scope is limited to an amount of material that can be covered in a one-semester course and the depth of presentation is restricted to the principal aspects of each model. The reference list directs the reader to sources that contain the additional details needed to produce more sophisticated models.

As the scientific and technological enterprise grows larger, scientists cope with the onslaught of new knowledge by retreating to more narrowly defined sub-disciplines. Increasing specialization tends to make each sub-discipline less relevant to the overall scientific and technological enterprise. This tempts scientists to pursue questions of trivial importance relative to the larger picture. Furthermore, the terminology that grows up in these enclaves becomes abstruse and incomprehensible. Geochemistry is challenged to avoid this trap by aggressively integrating knowledge from cognate disciplines in order to remain vibrant and meaningful. This is my justification for choosing many of the topics in this book from scientific and technological areas that might not be considered as traditional geochemistry. Hopefully this intellectual cross-fertilization will improve geochemistry's vigor. My choice of topics was predicated on the question: What should everyone know about rates of geochemical processes? I hope that this book answers that question. If I have done my job properly, this book will provide an intellectual foundation for anyone, including scientists and technologists from cognate disciplines, who deal with geochemical processes. I have tried to meld these topics by using a standard notation and terminology that is consistent with that already used in geochemistry.

I believe that understandable and reliable models are the basis of scientific knowledge, so this book emphasizes the development of conceptual and quantitative models. In addition, effective models make predictions that have practical value so these models are a valuable contribution to society. The essence of the scientific method is to compare a model's predictions with actual outcomes to test the validity of the underlying knowledge. Every scientist should understand and appreciate the importance of this model-building and -testing scenario.

This book does not delve into computer models of rate processes because there are already many good sources of information about this approach. Instead it focuses on fundamental and relatively simple models that are the foundations of the models incorporated into the various computer codes.

I have enjoyed the process of writing this book and learned a great deal from teaching this material to students. I hope that the readers will find it equally interesting and enlightening.

Chapter 1
Geochemical models

> *Modern geochemistry studies the distribution and amounts of the chemical elements in minerals, ores, rocks, soils, waters, and the atmosphere, and the circulation of the elements in nature, on the basis of the properties of their atoms and ions.*
>
> (Goldschmidt, 1958)

The distribution and circulation of the chemical elements in and on the Earth is influenced by a myriad of chemical and physical factors, many of which have changed over geological time. Understanding the role of these factors in geological processes requires us to condense information about elemental abundances and distributions into models. This book is about geochemical models for situations where time plays a key role. Geoscientists have always appreciated the importance of time in fashioning the Earth. Many geological processes require time spans that are far too long for human observation, but we can use models to extrapolate rates based on short-term observations to predict geochemistry in deep time. Equally important are models that forecast the future behavior of geochemical systems because those models are needed for environmental management and resources recovery projects.

Some of the models described in this book were developed by geochemists but many others come from applied sciences and engineering. Because of this diverse provenance, the models in their original form used a confusing mix of units, terminology, and notation. This book attempts to remedy that problem by recasting the models using internally consistent notation, units, and terminology familiar to geochemists. Furthermore, whenever possible the models are developed from fundamental theory showing a sufficient number of intermediate steps to allow the reader to follow the derivations.

Thermodynamic models have been a mainstay of geochemistry since the early twentieth century. They are especially effective for deep earth conditions where local equilibrium conditions prevail. However, at and near the Earth's surface extensive amounts of mass transport and low temperatures keep many reactions from reaching equilibrium. Kinetics models are needed to properly describe these situations. This makes the models of kinetic and dynamic processes described in this book complementary to thermodynamic models. The thermodynamic models describe the equilibrium endpoint for geochemical processes, and the kinetic and dynamic models predict how long it might take to reach that endpoint.

Kinetics and dynamics

Element redistribution in nature is the result of chemical reactions that are described by geochemical kinetics models and physical transport that is described by geochemical dynamics models. These geochemical kinetics and geochemical dynamics models are adapted from a diverse range of scientific and technological sources. Geochemical kinetics is grounded in well-established chemical kinetics theory (Laidler, 1987a, 1987b; Lasaga, 1998b) but that theory must be combined with practical methods to deal with the complexities of geological problems and technological applications. Valuable contributions to the field of geochemical kinetics and dynamics come from chemical engineering (Hill Jr., 1977; Levenspiel, 1972a), mineral processing (Burkin, 2001), civil and environmental engineering (Brezonik, 1994; Weber Jr. and DiGiano, 1996), soil science (Sparks, 1989), and even food processing (van Boekel, 2009). There are also several very useful books about mass transfer processes (Denny, 1993; Probstein, 1989; Vogel, 1994) that contain models that are readily adapted to deal with geochemical dynamics. Finally, there are many books that treat various aspects of geochemical kinetics (e.g. Berner, 1980; Lasaga, 1998b; Lerman, 1979; Stumm, 1990; Zhu and Anderson, 2002). Each of these sources has a particular scientific style that is reflected in different ways of making observations and constructing models. The challenge is to select relevant knowledge from these sources and to recast it in a way that makes it useful to geoscientists.

Model construction

Models organize our knowledge about the world. Doing science is the process of making observations, using those observations to develop a model, and then verifying the model's effectiveness by comparing its predictions with additional observations. The methods and scope of model-making vary from discipline to discipline, but the goal of creating reliable predictive tools is always the same. Developing a useful model requires a combination of creative thinking and disciplined use of modeling tools. There are many good discussions of modeling strategies (Aris, 1994; Bender, 1978; Bunge, 1997; Hall and Day, 1977; Hall *et al.*, 1977; Harte, 1988; Overton, 1977; Rescigno and Thakur, 1988; Shoemaker, 1977) but developing an effective model is in many ways a creative process. Good models are elegant and powerful. Elegance means that the model is expressed in the simplest, easiest to understand terms. Power means that the model explains the widest possible range of behaviors of the system of interest. Finally, good models are beautiful. Perhaps some of this beauty comes from power and elegance, but some of the beauty arises when the model is congruent with our thought processes. The esthetics of scientific creation is a difficult concept to learn (Wechsler, 1988), but it is a necessary part of making a lasting contribution to science.

Models can be categorized in various ways. Predictive models forecast the future behavior of a system, whereas conceptual models are used to understand relationships between system parts and processes. Deterministic models are constructed from mathematical functions that unambiguously relate cause and effect so that a particular set of input parameters produces a clearly related set of predicted results. Probabilistic models use statistical data to estimate the chance that an event or condition will occur. Forward models predict the future behavior of a system, whereas inverse (or reverse) models are used to extract fundamental data or mathematical relationships from past observations.

Models can relate information in three formats. Visualization illustrates spatial, mathematical, or sequential relationships using diagrams, graphs, and images. Narration uses words to explain how processes occur and are related to each other. Mathematics uses computational operations to quantitatively relate processes and features to each other. Disciplined use of notation, reliable data sources, and homogeneous units (Bender, 1978) along with appropriate mathematical methods and effective error analysis are features of effective mathematical models. However, it would be impossible to understand most mathematical models without accompanying visualization and narration. Visualization plays an equally important role in the development of scientific ideas (Wainer, 2005). For example, flow charts showing subsystems as boxes connected by arrows are often used to show how mass and/or energy moves through a system (McClamroch, 1980). This method of model construction, which is reprised in the ideal chemical reactor models described in Chapter 4, is especially useful because it leads to quantitative relationships in a simple and natural way. Graphs that allow us to visualize mathematical relationships are critical to understanding how dependent variables are influenced by independent variables. Harris (1999) provides extensive and detailed descriptions of many kinds of graphical formats. Effective graphical illustrations are easy to understand without needing to consult descriptive text. Effective graphs and illustrations maximize the information to ink ratio (Tufte, 2001). All models are built upon a conceptual foundation. Although visualizations often nicely summarize the conceptual basis of a model, it is usually impossible to fully explain the model without the use of some text. Well-organized and clearly stated narrative descriptions of the conceptual basis of a model are a key part of model development.

Model reliability

No model is perfect (Oreskes *et al.*, 1994). Even the best models fail under some conditions. In addition, making mistakes is a natural part of the modeling process. Systematic methods should be used to find, analyze, and correct mistakes and to define the valid range of a model. Determining the cause of a model's failure and repairing the model is a key task in model building.

Models are used to understand how complex behaviors arise from the interaction of simple processes. Ideally models are built upon reliable principles such as the conservation of matter, energy, and charge or the principle of detailed balancing. Often less-reliable relationships such as empirical rate equations must also be used. All quantitative models require input data and relationships that come from measurements, which always contain some error. This error can propagate through a model in unexpected ways, especially if the model simulates nonlinear interactions. This challenges the modeler who must decide whether an unexpected result is a legitimate prediction or simply an artifact arising from an unfortunate combination of errors. Three classes of errors occur in models. Formal errors are incorrect assumptions and/or formulations. They include errors in the conceptual foundation of the model as well as errors in the input data. Structural errors are errors in mathematical manipulations such as programming errors or algebraic errors. They include software bugs. Computational errors are errors in numbers caused by incorrect rounding or by addition or multiplication errors. Because there are so many ways that a model might fail, models and their predictions must be verified and validated to delineate the bounds of their reliability.

Verification tests whether the model is internally consistent, incorporates the correct relationships in the correct ways, and uses correct data. It is a good idea to develop a set of standards and practices that can be used in geochemistry model-making to insure the validity of the resulting model. Table 1.1 is an example of a checklist that might be used. Models should use a consistent system of notation and units to avoid structural errors. Equations should be tested using dimensional analysis. The input data should be reviewed to insure its correctness and internal consistency. Verification should determine the expected precision of the model's predictions based on the propagation of errors through the model. Error analysis is probably the most undervalued part of model development. For simple models, error propagation can be done using simple algebraic methods described in Chapter 2. For complex models, simple error propagation is often not practical. In such cases sensitivity analysis is a good strategy. Sensitivity analysis uses various schemes to systematically vary, within the expected range of error, the values of important modeling parameters to see how those variations affect the model's predictions. The utility of

Table 1.1. *Checklist of recommended validation and verification procedures*

Verification Questions
Are the equations used dimensionally homogeneous?
Are the standard and reference states of the variables internally consistent?
Is the notation internally consistent and clearly stated in the documentation for the model?
Are the computational methods correct? Have numerical simulations been tested against analytical solutions?
Are the chemical reactions correctly balanced? Are all the important chemical reactions and chemical species accounted for?
Does error analysis demonstrate that the predicted values are accurate enough to be useful?
Has the range and domain of the model been documented and is the situation being modeled within the range and domain?
Are the model's outputs clearly related to the inputs? Does sensitivity analysis show that each input has a significant effect on the model output?
Are the spatial and conceptual relationships consistent and properly documented?

Validation Questions
Are the predictions geologically reasonable? Are they consistent with reasonable estimates?
Are the predictions consistent with other scientific observations and knowledge?
Are the quantitative predictions sufficiently close to the behavior of one or more natural analogs?

a particular model depends upon whether it predicts the behavior of the system of interest within a practical range of uncertainty. For example, a model of a process used to control the lead content of drinking water would be useless if uncertainty in the predicted lead content was larger than the regulatory level. Typically models are constructed using very conservative assumptions that are frequently chosen on an arbitrary basis. Error analysis, on the other hand, clearly demonstrates quantitatively which of the input data most strongly affects the uncertainty of the predicted behavior.

Validation determines whether the model correctly predicts observable outcomes with reasonable accuracy and precision. Showing that a model can predict an analytical solution might be a useful validation test. Validation should compare a model's predictions with geological analogs. Models are hypotheses; they can never be proven true (Nordstrom, 2012). However, a model with a record of consistently successful predictions is more useful than an untested one. Finally, collecting new observations to test a model's predictions allows the model to be further refined. It is important to restate that the validating process does not prove that a model makes correct predictions; instead validation only decreases the probability of a model's failure.

Some technology models, usually related to resource recovery or waste disposal, involve predicting the behavior of the Earth after human intervention. Because there are frequently no precedents for such cases, models of the effects of human activities are sometimes validated using natural analogs, geological situations that are similar to the technological situation. In tests of technological models a clear definition of accuracy is needed. This definition is complicated because geochemical parameters almost always show a range of values, so comparison of a single model run with a single field observation has very little meaning. Instead, the input parameters for the model must be varied within the expected range to create a range of predictions that can be compared with the observed range of values. Then a statistical test can be used to estimate how often the model will fail.

> **Example 1.1.** Estimation method for validation of a chemical weathering denudation rate model
>
> Denudation is the lowering of the elevation of the Earth's surface by chemical weathering, erosion, and mass wasting. A model of chemical weathering would account for the rate of removal of material from soil and underlying rock by various dissolution processes. A first step toward validating such a model is to develop some independent estimates of the chemical denudation rate.
>
> One estimate might be based on geological knowledge. We know that denudation rates must be smaller than uplift rates. Otherwise the continents would be reduced to the elevation of sea level. The maximum uplift rate is probably about 10% of the maximum ocean ridge spreading rate, which is ~10 cm/yr. That makes uplift rates smaller than 1 cm/yr. The uplift rate of the average crust is likely to be 100 to 1000 times smaller than the maximum rate. We might choose the geometric mean of these values and estimate that the overall crust uplift rate is 0.03 mm/yr or 3 mm/century. A terrain with such a low uplift rate is likely to be relatively flat so mass wasting and erosion might account for two-thirds of the denudation, which means that chemical weathering would account for about one-third. This gives an estimate of denudation by chemical weathering of ~1 mm/century.
>
> Another way to estimate the chemical denudation rate is based on the realization that most of the soluble products of chemical weathering are carried away by infiltrating groundwater. If we consider 1 m^2 of land surface in a setting with about 100 cm of rainfall per year with 10% infiltration of the rainwater, ~100 L of water passes through each square meter annually. As that water infiltrates

into the water table, the concentration of dissolved species increases to perhaps 500 mg/L. The total mass of rock and soil dissolved away each year is 5×10^4 mg or ~50 g. If the density of the rock and soil minerals is ~3 g/cm^3, then 17 cm^3 (= 1.7×10^{-5} m^3) of material is carried away annually. If that volume is removed from under the 1 m^2 of land surface, the surface elevation will be lowered by 1.7×10^{-5} m (~0.02 mm). So the land surface elevation is reduced by 0.02 mm/yr or ~2 mm/century.

These estimates show that chemical weathering denudation rates should be on the order of a few millimeters per century. They might be as high as a few centimeters per century in humid tropical settings or as low as a few hundred microns per century in arid cold environments. Because these estimates constrain model predictions to fall within this range of values, they can be used as a validation test of a chemical weathering denudation model.

Harte (1988) and Weinstein and Adam (2008) provide detailed explanations about how to make meaningful estimations and they give a large number of practice examples.

Interpretation of results

The results of simple deterministic models are generally easy to interpret and apply. For example, a model of radioactive decay will predict that the radioactivity of a substance will be 1/8 of the original value after three half-lives have passed. As models become more complex they become more difficult to interpret. This suggests that model building should begin with highly idealized and simplified cases that are easily understood and tested. The next stage of model building involves testing the effect of potentially important variables on the model output to determine which variables are important enough to include and which can be neglected. This process of building models in a stepwise fashion not only eliminates a large number of unneeded independent variables but it helps the modeler develop a conceptual understanding of the situation being modeled.

The most important task of model building is to communicate expert knowledge developed by the model builder to others who can use that knowledge to solve problems. The modeling endpoint should be a report that contains thorough documentation of the model and an explanation of how to use its predictions. This report should contain a clear description of the model's conceptual basis in a well-written narrative section with accompanying illustrations and explanations of the mathematical methods. The report should illustrate relationships between the model's predictions and input parameters using response maps. These graphs of predictions versus parameters are much easier to understand than equations and tables.

Many of the figures in this book are response maps. The report should also describe the implementation of the modeling algorithm, including descriptions of special computational methods. The rationale for selecting the input data, including a discussion of their uncertainty, should be described. The range and domain of the model should be specified. Finally, the report should explain how to interpret the model's output.

Chapter 2
Modeling tools

Before any model is ready for use it must be verified. The verification step is greatly simplified if the model is constructed using conventional computational methods, notation, and units. This chapter reviews some procedures and conventions that are recommended for geochemical model construction.

Balancing chemical reactions

The first step in building a geochemical model is to write balanced equations that describe the governing chemical reactions. It is often possible to recognize these reactions based on past experiences, but when experience is lacking a general strategy is needed to identify these key reactions. The strategy should recognize that, with few exceptions, the key reactions involve the most abundant phases and chemical species. Creating a mineral inventory listing the possible hosts for the elements of interest is a first step toward selecting the solid phases to include in the model. Similarly, a chemical analysis of the aqueous phase can be used along with an aqueous speciation model to identify important aqueous species. The reactions among these mineral and aqueous species are expressed as balanced chemical reactions and these reactions become the basis for the subsequent model.

Once the key solid phases and aqueous species have been recognized, the next step is to write the balanced chemical reactions that link them. Most geochemical reactions can be balanced by inspection, which involves the stepwise selection and adjustment of tentative reaction coefficients. All reaction balancing is based on the concept of conserved atomic identity, which means that reactions only change the associations among the atoms without changing their atomic identity. This means that there must be an equal number of each kind of atom on both sides of the reaction. Furthermore, the conservation of charge requires that the total charge must be the same on both sides of a balanced chemical equation. These requirements are enforced in five steps.

1. Write the chemical formulae for the reactant and product species on opposite sides of an equal sign.
2. Find the element that is least represented in these species and insert integer (or sometimes small fraction) coefficients into the equation to make equivalent amounts of this element on each side of the equation.
3. Repeat this step with the second least represented element and so on until all the elements are balanced. For reactions involving aqueous solutions, it is usually necessary to add H_2O, O_2, H_2, or H^+ to balance the hydrogen and oxygen.
4. Adjust the coefficients of the charged species to produce an equivalent amount of charges on each side of the equation.
5. Finally, divide the entire equation through by integers to reduce these coefficients to small integer values (although sometimes small fractions are acceptable).

Although balancing equations by inspection is the most convenient method, the process can be daunting for reactions involving complicated molecules. For the more challenging reactions, various algebraic methods are recommended. The reaction coefficients are related by a series of linear equations, so complex reactions can be balanced by solving these equations either by the substitution method or by using matrix algebra (Campanario, 1995; Krishnamurthy, 1978; Presnall, 1986). The various matrix methods are well suited for incorporation into computer codes.

Example 2.1. Pyrite oxidation reaction

Acid mine drainage is a common environmental problem associated with disposed mine wastes that contain iron sulfide minerals, especially pyrite. Acid mine drainage is a low pH solution containing relatively high concentrations of dissolved iron and sulfate. Acid mine drainage streams are stained yellow, brown, and red by precipitated ferric oxyhydroxides and hydroxysulfates. We know that oxidation of pyrite (FeS_2) to produce sulfuric acid is the primary cause of acid mine drainage and that, although there are often several different ferric oxyhydroxides and hydroxysulfate minerals present, goethite (FeOOH) is one of the most important. Therefore, constructing a model of acid mine drainage geochemistry might begin by developing a chemical equation that describes the overall acid mine drainage process.

1. We know that during the formation of acid mine drainage, pyrite is destroyed and goethite is formed so the first step is to put pyrite on the left side of the equation and goethite and sulfate on the right.

$$FeS_2 = FeOOH + SO_4^{2-}$$

2. Inspection of this unbalanced "equation" shows that an equal number of iron atoms occur on each side so no action is needed to balance the iron.

$$FeS_2 = FeOOH$$

3. However, there must be two sulfates on the right-hand side to balance the two sulfur atoms in the pyrite:
$$FeS_2 = FeOOH + 2\ SO_4^{2-}$$

4. Next, there must be 0.5 H_2O on the left-hand side to balance the H in the goethite on the right-hand side.
$$FeS_2 + 0.5\ H_2O = FeOOH + 2\ SO_4^{2-}$$

5. Because the sulfur in the pyrite is reduced, with a formal charge of -1, and the sulfur in the sulfate is oxidized, with a formal charge of $+6$, we know that an oxidant, in this case O_2, must occur on the left-hand side of the reaction. There are already 10 oxygen atoms on the right-hand side of the reaction and only 0.5 on the left so 9.5 oxygen atoms (4.75 O_2) are needed on the left-hand side.
$$FeS_2 + 4.75\ O_2 + 0.5\ H_2O = FeOOH + 2\ SO_4^{2-}$$

6. Finally, four hydrogen ions are needed on the right-hand side to charge balance the sulfate. These can come from the addition of two more water molecules to the left-hand side.
$$FeS_2 + 4.75\ O_2 + 2.5\ H_2O = FeOOH + 4\ H^+ + 2\ SO_4^{2-}$$

However, this adds too much oxygen to the left-hand side, so one O_2 must be removed.
$$FeS_2 + 3.75\ O_2 + 2.5\ H_2O = FeOOH + 4\ H^+ + 2\ SO_4^{2-}$$

7. This reaction is often written with decimal fractions as shown above but the coefficients can be written as simple fractions.
$$FeS_2 + 15/4\ O_2 + 5/2\ H_2O = FeOOH + 4\ H^+ + 2\ SO_4^{2-}$$

or the equation can be multiplied by 4 to get
$$4\ FeS_2 + 15\ O_2 + 10\ H_2O = 4\ FeOOH + 16\ H^+ + 8\ SO_4^{2-}$$

A more systematic alternative to this approach begins by expressing the reaction coefficients as algebraic variables and then solving the resulting equations by the substitution method.
$$a\ FeS_2 + b\ O_2 + c\ H_2O = d\ FeOOH + e\ H^+ + f\ SO_4^{2-}$$

Then the relationships between the coefficients that are required by the conservation of mass and charge are expressed by simple equations.

Fe: $a = d$
S: $2a = f$
O: $2b + c = 2d + 4f$
H: $2c = d + e$
chg: $e = 2f$

> These equations are solved in a stepwise fashion to find the numerical value for each coefficient.
>
> Let $a = 1$.
> From Fe: $\quad d = a = 1$.
> From S: $\quad f = 2a = 2$.
> From chg: $\quad e = 2f = 4$.
> From H: $\quad 2c = 1 + 4$ so $c = 5/2$
> From O: $\quad 2b = 2d + 4f - c = 2 + 8 - 5/2 = 15/2$ so $b = 15/4$
>
> These coefficients produce a balanced equation.
>
> $$1\ FeS_2 + 15/4\ O_2 + 5/2\ H_2O = 1\ FeOOH + 4\ H^+ + 2\ SO_4^{2-}$$
>
> In all cases, the value for one of the coefficients must be arbitrarily chosen. The other coefficients are then scaled to that value.

Notation

The notation used in this book follows geochemistry conventions as closely as possible but has been modified where necessary to avoid redundancy or confusion. Geochemical modeling requires an easy to recognize and internally consistent set of notation. Inconsistent notation complicates model verification and leads to errors. The notation used in geochemistry is often different from that used in engineering, chemistry, and biology, and sometimes models require the introduction of new notation. In all cases, notation should be clearly defined in the model's documentation and it is good practice to include a table of notation, with units, in every report.

Thermodynamic functions, especially free energy, enthalpy, and entropy, are often used in kinetics models. Because thermodynamics is so widely used in science, many different kinds of notation have developed. The convention chosen for this book is illustrated using Gibbs free energy as an example. For the free energy change related to a reaction, the delta in front of ΔG_r° means that this variable reflects the change in free energy in going from the reactants to the products. The subscript, $_r$, designates this variable as the free energy of reaction. The superscript, $^\circ$, indicates that both the reactants and the products are in the standard state, 298.15 K (25°C) and either 1 atm or 100 kPa (1 bar) total pressure. For the free energy of formation of a substance from the elements, the subscript, f, associated with ΔG_f° means that this is the free energy related to the formation from the elements (this is essentially the same as ΔG_r° except that the free energy of formation of the elements in their most stable state at 298.15 K (25°C) and either 1 atm or 1 bar total pressure is defined as zero. It is often useful to use an appended notation for $\Delta G_f°$ and $\Delta G_r°$ to indicate what reaction or substance the variable refers to. For example, the free energy of reaction for a reaction listed in equation (1) would be $\Delta G_r°(1)$ and the free energy of formation of quartz would be $\Delta G_f°(qz)$.

Concentration and activity notation can be very confusing. Chemists have long used square brackets, [i], to indicate the activity of species i and parentheses, (i), to indicate the concentration of species i, but geochemists have occasionally reversed this convention. To make things worse, curly brackets are used in some engineering disciplines to indicate activity. This confusion is best avoided by abandoning the bracket-parenthesis notation in favor of using a_i to denote activity and m_i to indicate molal concentration. Although molar concentration units are best avoided in geochemistry, when used they can be represented as M_i. Concentrations given other units such as mol/m³, ppm, or mg/L units are best represented as c_i.

pX notation indicates the negative logarithm (base 10) of quantity X. Most people are familiar with pH, which is the negative logarithm of the hydrogen ion activity, i.e. pH = $-\log a_{H^+}$. By analogy pK = $-\log K$ and pCl = $-\log a_{Cl^-}$.

Kinetics variable notation is quite different from discipline to discipline and often inconsistent within a single discipline. In this book, r_i refers to the change in the number of moles of species i in a system per unit time (mol/sec), R_i refers to the change in concentration of species i per unit time (molal/sec), and J_i refers to the flux of species across an interface (mol/m²sec).

Units

The International System of Units was devised to facilitate communication between scientific and engineering disciplines. Using SI units is highly recommended because it minimizes errors that arise from complicated units conversions. The International System of Units is based on seven base units (Table 2.1). Decimal multiples of these units and derived units are named using the prefixes listed in Table 2.2.

Concentration

Concentration units listed in Table 2.3 are frequently used by geochemists. Thermodynamic calculations are typically based on molal concentration and concentration should be expressed in molal units whenever possible. Some disciplines use molar or millimolar concentration units supposedly because of the convenience of mixing of solutions using volumetric flasks. Unfortunately molar concentrations change whenever the solution density changes because of changing temperature, pressure, or composition. This

Table 2.1. *Base units and important related units in the SI*

Quantity	Name	Symbol
SI base units		
length	meter	m
mass	gram	g
time	second	s
electric current	ampere	A
temperature (thermodynamic)	kelvin	K
amount of substance	mole	mol
luminous intensity	candela	cd
Derived units		
volume	liter	$1 \text{ L} = 10^{-3} \text{ m}^3$
force	newton	$1 \text{ N} = 1 \text{ kg m}^{-1} \text{ sec}^{-2}$
energy	joule	$1 \text{ J} = 1 \text{ N m}$
	calorie[1]	$1 \text{ cal} = 4.184 \text{ J}$
pressure	pascal	$1 \text{ Pa} = 1 \text{ N m}^{-2}$
	bar	$1 \text{ bar} = 10^5 \text{ Pa}$
	atmosphere	$1 \text{ atm} = 101325 \text{ Pa}$
power	watt	W
electrical charge	coulomb	C
electrical potential	volt	V

[1] Widely used but not recommended.

Table 2.2. *Decimal multiples of SI units*

Factor	Prefix	Symbol	Factor	Prefix	Symbol
10^{24}	yotta	Y	10^{-1}	deci	d
10^{21}	zetta	Z	10^{-2}	centi	c
10^{18}	exa	E	10^{-3}	milli	m
10^{15}	peta	P	10^{-6}	micro	μ
10^{12}	tera	T	10^{-9}	nano	n
10^{9}	giga	G	10^{-12}	pico	p
10^{6}	mega	M	10^{-15}	femto	f
10^{3}	kilo	k	10^{-18}	atto	a
10^{2}	hecto	h	10^{-21}	zepto	z
10^{1}	deka	da	10^{-24}	yocto	y

Table 2.3. *The most commonly used concentration units for aqueous solutions*

Unit	Definition
milligrams per liter, mg/L	milligrams solute/1 L solution
millimolar, mM	moles solute/10^3 L solution
molal, m	moles solute/1 kg water
molar, M	moles solute/1 L solution
mole fraction, X	moles solute/Σ(moles of all solutes and solvents)
parts per billion, ppb	grams solute/10^9 grams solution
parts per million, ppm	grams solute/10^6 grams solution
per mil, ‰	grams solute/10^3 grams solution

means that a complex transformation is needed to make molar concentration consistent with thermodynamic standard states (mole fraction or molal scales). Molar and molal concentrations are nearly equivalent when the solutions are dilute (<~0.1 M or m) and near 25°C because the solution density is near 1 g/cm^3.

Some conventions for reporting concentrations of aqueous species are often confusing to the uninitiated. Sometimes conventions are chosen because they make calculations convenient but some are based on historical practice. To avoid confusion, it is useful to follow the most widely used conventions. However, there are cases when these conventions should be violated to simplify a calculation. Whenever conventions are violated the new usage should be carefully explained in the documentation that accompanies the calculations and models. A good example of a convention that makes calculations easier is the case of dissolved CO_2. Carbon dioxide dissolves in water via two different reactions.

$$CO_2(v) = CO_2(aq) \tag{2.1}$$

$$CO_2(v) + H_2O = H_2CO_3(aq) \tag{2.2}$$

This means that both $CO_2(aq)$ and $H_2CO_3(aq)$ are present in the solution. Because there is no simple way to determine the concentration of these individual species, the concentration of unionized, dissolved CO_2 is usually reported as the sum of the concentrations of $CO_2(aq)$ and $H_2CO_3(aq)$. This convention is quite useful for thermodynamic calculations and the reported equilibrium constants for reactions involving $H_2CO_3(aq)$ as well as the $\Delta G_f^\circ(H_2CO_3(aq))$ are based on the convention of lumping these two species together. However, this convention may cause a problem in kinetics models because the rate of reaction of $CO_2(aq)$ is likely to be different from the rate of reaction of $H_2CO_3(aq)$.

A convention that is based on historical practice is reporting dissolved silica concentrations as mg SiO_2/L or as ppm SiO_2 rather than in terms of Si or H_4SiO_4. This convention appears to be a relic from the time when the concentration of elements in minerals was reported in terms of oxide components, i.e. weight percent oxides.

There are several properties of aqueous systems that are described using well-defined operations. The variables derived from these operations can be used in rate models, but their usefulness is often restricted because of their incomplete link to fundamental (SI) units. Most of the important operational methods are carefully laid out in Greenberg *et al.* (1992) and their significance is explained in Hem (1985). Sometimes comparable values can be derived by equivalent operations; for example, hardness is now determined by measuring the calcium and magnesium concentrations of solutions rather than by using a soap test. Although operationally defined values often have important technological uses, they are seldom useful for detailed modeling of the thermodynamic or kinetic reactivity of the solution but they often can be converted into variables that are useful for modeling.

Acidity is measured by titrating the sample with a NaOH solution to raise its pH to a defined endpoint (usually 8.3). This measures not only the concentration of free hydrogen ions in the sample but also the amount of hydrogen ions that are liberated by various hydrolysis reactions. Acidity is reported as mg $CaCO_3$/L equivalent. This can be interpreted as the amount of calcite that would have to be added to the solution to raise the pH to 8.3. Solutions like acid mine drainage or oil field brines contain reduced redox sensitive species (e.g. Fe^{2+}, H_2S) that upon oxidation and hydrolysis produce hydrogen ions. These solutions are treated with hydrogen peroxide before the titration to release this latent acidity. Kirby and Cravotta III (2005a) and Kirby and Cravotta III (2005b) provide a detailed explanation of latent acidity in acid mine drainage.

Alkalinity is a measure of the acid-neutralizing capacity of the solution. It is measured by titrating the solution to pH 8.3 to determine phenolphthalein alkalinity and to a pH near 4.5 (see Greenberg *et al.* (1992) for exact values) to determine total alkalinity. Note that most values reported as alkalinity are total alkalinity. Alkalinity is also reported as mg $CaCO_3$/L equivalent, which can be interpreted as the equivalent amount of calcite needed to consume the amount of acid titrated. Because bicarbonate is usually the predominate anion in most non-marine surface waters and shallow groundwater, total alkalinity values are sometimes interpreted as a reflection of the bicarbonate concentration of the solution. This interpretation should be applied with caution because several other kinds of anions, including hydroxide, organic acids, phosphates, silicates, carbonate, and borate, contribute to the measured alkalinity.

Conductivity is a measure of the ability of an aqueous solution to carry an electric current between two chemically inert electrodes. To avoid polarization of the electrodes, an alternating current signal is used. Because the current is carried by ions, conductivity is related to the amount of dissolved electrolytes so that the salinity of the solution can be estimated from the measured conductivity (see Greenberg *et al.* (1992)).

Eh measures the electrical potential difference, in volts, between an inert electrode (usually Pt) and a reference electrode (usually Ag-AgCl) that has been calibrated relative to the standard hydrogen electrode using ZoBell's solution (Nordstrom, 1977). This value is frequently not equivalent to thermodynamic Eh values because the slow rates of many redox reactions keep them from attaining equilibrium at the Pt electrode.

Hardness was originally designed to measure the capacity of the water to react with and precipitate a standard soap. Although other cations, including H^+ and Fe^{2+}, can react with soaps, hardness is currently defined as the sum of the Ca^{2+} and Mg^{2+} concentrations because these cations are usually the most abundant ones in natural waters. Water hardness is reported as mg $CaCO_3$/L equivalent and this quantity is calculated as

$$\text{mg } CaCO_3/L = 2.5 \text{ (mg/L) Ca} + 4.1 \text{ (mg/L) Mg} \tag{2.3}$$

Salinity is a measure of the mass of dissolved salts in the solution. This quantity is commonly estimated using an empirical relation between some measured physical quantity, like conductivity, refractive index, sound speed, or density, and the dissolved salt content of a standard solution. The currently recommended methods to determine salinity use electrical conductivity or density. Details of the methods are given in Greenberg *et al.* (1992). Salinity is reported as a dimensionless number such as % or ‰.

TDS is a measure of the total dissolved solids, including unionized species like SiO_2, in the solution. This measurement is slightly different from salinity, which measures only the amount of ionizable salts. TDS is determined from the weight of the residue of a filtered solution evaporated to dryness at 180°C. It is reported in units of mg total dissolved solids/L.

Dimensional analysis

A dimension is a quantity such as time, mass, temperature, or distance, that can be expressed numerically in terms of one or more standard units. Dimensions have homogeneous linear scales so that a 2-meter interval measured at the beginning of a 1-kilometer traverse is identical in length to a 2-meter interval measured near the end of the traverse. The magnitude of a dimension is expressed in terms of a pure number and a unit of measure. Units are standards to which dimensional quantities are compared to obtain a numerical ratio, which is the pure number. For example, comparison of a football field to a yardstick gives the ratio of 100:1 allowing us to say that the football field is 100 units long and the unit of measure is a yard which has a universally agreed-upon standard length. This means that we can define any unit of measure for a dimension as long as it is related to any other unit of measure for that dimension by a constant ratio. This allows us to convert dimensional measurements from one set of units to another using the method of unit conversions discussed below. The concept of dimension requires that mathematical models of dimensional systems have homogeneous dimensions so that the individual terms of the defining equations must have the same units. The methods of dimensional analysis and similitude are widely used in engineering and physics to identify relationships between variables used in quantitative models and to argue that these relationships should apply over a wide range of scales. When nonlinear relationships between variables occur, the variables must be expressed in terms of dimensionless numbers to avoid violating the principle of similitude. There are many books about dimensional analysis but most of the following was drawn from Douglas (1969), Schepartz (1980), and Zlokarnik (1991).

One of the simplest applications of dimensional analysis is to convert a measured or calculated dimension from one system of units to another. This simply involves multiplying the quantity by conversion factors until all the units cancel out except the desired ones. For example, if a solution contains 80 ppm Ca^{2+}, the molal concentration is found by the following unit conversions:

$$80 \text{ ppm} \approx \frac{80 \text{ mg}}{1 \text{ kg}} \times \frac{1 \text{ g}}{1000 \text{ mg}} \times \frac{1 \text{ mol}}{40 \text{ g}} = 2 \times 10^{-3} \frac{\text{mol}}{\text{kg}} = 2 \times 10^{-3} \text{ m} \qquad (2.4)$$

The universal gas constant can be converted from volume and pressure units to energy units.

$$\frac{8.314 \times 10^{-2} \text{ dm}^3 \text{ bar}}{\text{mol K}} \times \frac{1000 \text{ cm}^3}{1 \text{ dm}^3} \times \frac{1 \text{ J/bar}}{10 \text{ cm}^3} = \frac{8.314 \text{ J}}{\text{mol K}} \qquad (2.5)$$

A fundamental principle of dimensional analysis is that after all unit conversions are completed the same units must appear on the right- and left-hand side of an equation. This principle of homogeneous units is an extremely useful way to verify a calculation. If the terms of an equation do not have homogeneous units then the equation is incorrect. Even though homogeneous units are not a complete guarantee of the correctness of an equation, passing this test does give substantial confidence in the equation.

Because geochemical processes are often quite complex, situations for which no quantitative model exists are often encountered. These situations can be handled by developing a model using dimensional analysis.

This approach assumes that the behavior of the system is unrelated to its scale. For example, we assume that if the reaction between Ca^{2+} and HCO_3^- forms calcite in 100 g of solution in a laboratory beaker containing synthetic sea water, it will form calcite in exactly the same way in the vast reaches of the ocean as long as the solution concentrations are the same. This idea of scale invariant behavior of systems is called similitude. Principles of similitude have long been used to build physics and engineering models (Bridgman, 1931; Huntley, 1967; Kline, 1965; Murphy, 1950; Staicu, 1982). Although this principle is very powerful and the resulting models are typically valid over orders of magnitude of scale, quantum effects can cause similitude-based models to fail at scales approaching the size of atoms.

Example 2.2. Surface area to volume ratio of a sphere

The surface area to volume ratio of a sphere is used in several models in this book. Dimensional analysis can be used to find the relationship between the surface area (A) with dimensions of $[L^2]$ and the internal volume (V) with dimensions of $[L^3]$ for any convex regular solid. The surface area is related to the volume by a dimensionless constant, b.

$$[L^2] = b[L^3]^{2/3} \tag{2.6}$$

$$A = bV^{2/3} \tag{2.7}$$

The first step toward finding the value of b, substitutes the definitions of A and V for a sphere into Eq. (2.7).

$$4\pi r^2 = b\left(\frac{4}{3}\pi r^3\right)^{2/3} \tag{2.8}$$

Equation (2.8) can be solved to find b.

$$b = \frac{4\pi}{\left(\frac{4}{3}\right)^{2/3} \pi^{2/3}} = \frac{4\pi^{1/3}}{\left(\frac{4}{3}\right)^{2/3}} = 4.84 \tag{2.9}$$

For cubes, $b = 6$. Other regular convex solids have intermediate values of b. Note that when this relationship between area and volume is scaled downward, we eventually find a situation where solids do not occupy space in a way that is compatible with this model. That is because the actual dimensions of an atom are best expressed in terms of probability density functions rather than as absolutely fixed spatial dimensions.

The concept of similitude is further formalized by the Buckingham π theorem that states that an equation containing n variables can be rewritten as an equation containing $n - m$ dimensionless numbers, where m is the overall number of independent dimensions in the model.

Dimensionless numbers are needed whenever nonlinear equations are encountered because transcendental and polynomial functions cannot have units. Taking the logarithm of 10 moles or raising e to the power of 20 minutes is a meaningless calculation. This does not mean that there cannot be units within these functions but it does mean that those units must cancel out. For example, it is perfectly acceptable to raise e to kt where t is time and the rate constant, k, has the dimension of reciprocal time so that the product kt has no units. Recasting a variable into a dimensionless number for use in a nonlinear equation can be done in various ways, but all those ways are subject to the restriction of the Buckingham π theorem.

Logarithms

Logarithmic transformations are quite common in geochemical models. The variables used in geochemical calculations often range over many orders of magnitude so that it is more convenient to work with the logarithms of these numbers rather than the numbers themselves. In addition, many relationships in thermodynamics and kinetics are linearized using logarithmic transformations.

A logarithm is the power (p) to which a base (a) must be raised to produce a number (N).

$$a^p = N \qquad (2.10)$$

Adding logarithms is equivalent to multiplying numbers so it is often convenient to express numbers as powers of ten. Adding the exponents is equivalent to multiplying the numbers.

$$245 \times 63 = (10^{2.39})(10^{1.80}) = 10^{2.39+1.80} = 10^{4.19} = 1.6 \times 10^4$$

Likewise division is accomplished by subtracting powers of ten.

$$245/63 = 10^{2.39}/10^{1.80} = 10^{2.39-1.80} = 10^{0.59} = 3.9$$

Base 10 logarithms are so often used in geochemical calculations that it is helpful to memorize approximate values for the logarithms of the numbers 1 through 9. The following table offers some memory aids.

log 1 = 0.0	10^0
log 2 = 0.301 ≈ 0.3	Memorize
log 3 = 0.477 ≈ 0.5	Memorize
log 4 = 0.602 ≈ 0.6	$= 2^2 =$ 2 log(2)
log 5 = 0.699 ≈ 0.7	Halfway between 4 and 6
log 6 = 0.778 ≈ 0.8	= (2)(3) = log(2) + log(3)
log 7 = 0.845 ≈ 0.85	Halfway between 6 and 8
log 8 = 0.903 ≈ 0.9	$= 2^3 =$ 3 log(2)
log 9 = 0.954 ≈ 0.95	Halfway between 8 and 10
log 10 = 1.0	10^1

Note that the largest errors associated with this method are for the logarithms of 3 and 6, both of which are incorrect by ~0.02 log units.

Errors

There is some uncertainty in all data, and model building must take this error into account. The first step in error management is error detection, error reduction, and error quantification. There are three types of error: systematic error, random error, and blunders. Improved experimental protocol can reduce all these, but designing progressively better experiments eventually leads to diminishing returns so that at some point it is necessary to use some kind of error analysis to manage the uncertainty in the variable being quantified.

Systematic errors affect the accuracy but not the precision of the result. They are usually errors in calibration or observation where the same incorrect protocol is applied to all measurements. They displace all measurements from the true value by the same amount so they cannot be detected by a statistical analysis of only one data set. However, systematic error can be detected and reduced by comparing data sets from several different sources using meta-analysis and systematic review (Rimstidt *et al.*, 2012).

Random error displaces individual measurements from the true value, but as the number of determinations increases their average becomes closer to the true value. In a series of independent measurements that contain only random errors, the most reliable estimate of the true value is the mean of all the measurements. So that for a series of N measurements, $x_1, x_2, x_3, \ldots x_N$, the best estimate of the true value of x is $\bar{x}$.

$$\bar{x} = \frac{\sum_{i=1}^{n} x_i}{N} \qquad (2.11)$$

The standard deviation of the measurements, σ_x, is a measure of the amount of random error associated with the measurements but it is not a measure of the error associated with $\bar{x}$.

$$\sigma_x = \sqrt{\frac{\sum(x_i - \bar{x})^2}{N-1}} \qquad (2.12)$$

The best estimate of the uncertainty in $\bar{x}$ is the standard error of the mean, $\sigma_{\bar{x}}$.

$$\sigma_{\bar{x}} = \frac{\sigma_x}{\sqrt{N}} \qquad (2.13)$$

This uncertainty is reported as $\bar{x} \pm \sigma_{\bar{x}}$ or as $\bar{x}(\sigma_{\bar{x}})$. The latter convention is used in this book.

Blunders are one-time mistakes that are random but not normally distributed. They might be the result of misreading an instrument or an incorrect notebook entry. The most egregious blunders can be identified and rejected from the data set using Chauvenet's criterion. Chauvenet's criterion is based on the idea that a typical set of experimental data will have a normal distribution of errors (Taylor, 1982). That is, 68.3% of the data will fall within one standard deviation of the mean; 95.5% will fall within two standard deviations; and 99.7% will fall within three standard deviations. If the data set contains values that do not conform to this pattern, then these values are likely to be the result of some error(s) other than the expected random errors that produce a normal distribution. Because these unusual values are not part of the expected pattern they should be discarded from the data set. Keep in mind that this approach assumes a normal distribution of error. If the measurement produces a different kind of error distribution, Chauvenet's criterion cannot be used.

The procedure for applying Chauvenet's criterion to a series of measurements, $x_1, x_2, x_3, \ldots x_N$, is as follows:

1. Calculate the mean and standard deviation of all the data.
2. Select a suspicious datum (x_s) and calculate the number of standard deviations (t_s) between the point and the mean.

$$t_s = \frac{x_s - \bar{x}}{\sigma_s} \qquad (2.14)$$

3. Use a probability table to determine the probability (P_s) that a normally distributed measurement will differ from the mean by t_s standard deviations ($P_s = 1 - P_{table}$).

4. Determine the number of measurements in the data set that are expected to be as bad as x_s by multiplying the total number of data (N) by the probability (P_s).

$$N_s = NP_s \qquad (2.15)$$

5. If $N_s < 0.5$, then x_s fails Chauvenet's criterion and is rejected along with all data that differ from the mean by more than x_s.
6. After the data are rejected, calculate a new mean and uncertainty using the remaining data.
7. Chauvenet's criterion can be applied to a data set only one time.

Significant digits

The significant digits in a number give an indication of the uncertainty associated with that value. This means that the number has been rounded to the same order of magnitude as its uncertainty. The number of significant digits is determined by the following rules:

1. The leftmost nonzero digit is the most significant one.
2. In numbers with no decimal point, the rightmost nonzero digit is the least significant one.
3. In numbers with a decimal point, the rightmost digit is the least significant one, even if it is a 0.
4. All digits between the least and most significant are counted as significant digits.

For example, 65000 has two significant digits; 0.1010 has four significant digits; and 65000.1010 has nine significant digits. When numbers are truncated to the proper number of significant digits, the least significant digit must be rounded up or down according to the following rules:

1. If the digit following the one to be retained is 6, 7, 8, or 9, round the retained digit up by one.
2. If the digit following the one to be retained is 0, 1, 2, 3, or 4, simply drop it without changing the retained digit.
3. If the digit is 5 and the digit to be retained is even, do not change it. If it is odd, round it up by one.

The simplest method of error propagation involves keeping track of significant digits during a calculation and rounding the answer to the same number of significant digits as occurs in the least significant number used for the calculation. This procedure gives a very conservative account of the possible error in a calculated value. As a rule of thumb, the number of significant digits in a reported value should not exceed the smallest number of significant digits in any number used to calculate that value. For a number with a reported uncertainty, the least significant digit in the stated value should be of the same order of magnitude (in the same decimal position) as the most significant digit of the uncertainty. It is customary to perform the calculation using numbers with additional significant digits, to avoid introducing rounding errors and then rounding the final answer to the correct number of significant digits. For a number without a reported uncertainty, one should assume that all reported digits are significant. Geochemical measurements typically have more than 1% error so that, if no uncertainty is stated, three significant digits are likely to be sufficient to express the reliable part of the number.

Propagation of errors

Precision is lost as variables are manipulated through formulae. This loss of precision is taken into account by propagating the uncertainty associated with the model's variables through the equations. There are several standard references that provide detailed discussions of error propagation and management (Bevington, 1969; Deming, 1943; Mandel, 1964; Taylor, 1982).

For the propagation of uncertainties in equations with addition or subtraction, the uncertainty in the calculated value is the square root of the sum of the squares of the absolute uncertainties.

$$q = x + y + z \cdots a - b - c \qquad (2.16)$$

$$\delta q = \sqrt{(\delta x)^2 + (\delta y)^2 + (\delta z)^2 + (\delta a)^2 + (\delta b)^2 + (\delta c)^2} \qquad (2.17)$$

For the propagation of uncertainties in multiplication or division, the fractional uncertainty in the calculated value is the square root of the sum of the squares of the fractional uncertainties.

$$q = \frac{xyz}{abc} \qquad (2.18)$$

$$\frac{\delta q}{q} = \sqrt{\left(\frac{\delta x}{|x|}\right)^2 + \left(\frac{\delta y}{|y|}\right)^2 + \left(\frac{\delta z}{|z|}\right)^2 + \left(\frac{\delta a}{|a|}\right)^2 + \left(\frac{\delta b}{|b|}\right)^2 + \left(\frac{\delta c}{|c|}\right)^2} \qquad (2.19)$$

For the propagation of uncertainty where a constant is known exactly, the uncertainty in the calculated value is the product of the constant and the uncertainty in the variable.

$$q = Bx \qquad (2.20)$$

$$\delta q = |B|\delta x \qquad (2.21)$$

For the propagation of uncertainties through a function of one variable, the uncertainty in the calculated result is the product of the derivative of the function and the derivative of the variable.

$$q = q(x) \qquad (2.22)$$

$$\delta q = \left|\frac{dq}{dx}\right|\delta x \qquad (2.23)$$

Conversion between logarithms and antilogarithms is so common in geochemical models that it is worth showing the error propagation for these special cases of the application of Eq. (2.23). The antilog transformation converts $\log k$ to k so $q(x) = 10^x$.

$$dq = \left|(10^x)(\ln 10)\right|\delta x \qquad (2.24)$$

$$\delta x = 2.303(10^{(\log x)})\delta_{\log x} \qquad (2.25)$$

The logarithmic transformation converts k to $\log k$ so $q(x) = \log(x)$.

$$\delta q = \left|\frac{\log e}{x}\right|\delta x \qquad (2.26)$$

$$\delta(\log x) = \left|\frac{0.434}{x}\right|\delta x = 0.434\left|\frac{\delta x}{x}\right| \qquad (2.27)$$

The last term in this equation is the fractional error in x.

For power law functions, the fractional uncertainty is the product of the absolute value of the exponent and the fractional uncertainty in the variable.

$$q = x^n \qquad (2.28)$$

$$\frac{\delta q}{|q|} = |n|\frac{\delta x}{|x|} \qquad (2.29)$$

In the most general case, the uncertainty in the dependent variable of a function with several independent variables is given by Eq. (2.31).

$$q = f(x, y, ... z) \qquad (2.30)$$

$$\delta q = \sqrt{\left(\frac{\partial q}{\partial x}\delta x\right)^2 + \left(\frac{\partial q}{\partial y}\delta y\right)^2 + \cdots + \left(\frac{\partial q}{\partial z}\delta z\right)^2} \quad (2.31)$$

In all cases, the maximum fractional uncertainty in a calculated result is never larger than the sum of the fractional uncertainties in the variables.

$$\frac{\delta q}{|q|} \leq \frac{\delta x}{|x|} + \frac{\delta y}{|y|} + \cdots + \frac{\delta z}{|z|} \quad (2.32)$$

Example 2.3. Some examples of error propagation

Addition and/or subtraction: It is sometimes necessary to composite samples in order to have enough material for a chemical analysis. If three samples with masses of 251 mg, 172 mg, and 302 mg that were weighed on a balance with an error of 2 mg are composited, the total mass of sample is 725 mg and the uncertainty associated with this value is found using Eq. (2.17).

$$\delta q = \sqrt{2^2 + 2^2 + 2^2} = 3.46 \quad (2.33)$$

This means that the mass of the composited sample is 725(3.46) mg.

Multiplication and/or division: The rate of reaction in a mixed flow reaction is calculated by multiplying the concentration (mol/g) of a chemical species in the effluent solution by the flow rate (g/sec). If the concentration is $2.31(0.0122) \times 10^{-4}$ mol/g and the flow rate is 0.124(0.00131) g/sec, the rate is 1.53×10^{-5} mol/sec and the uncertainty associated with this value is found using (2.19).

$$\frac{\delta q}{q} = \sqrt{\left(\frac{1.22 \times 10^{-6}}{|2.31 \times 10^{-4}|}\right)^2 + \left(\frac{0.00131}{|0.124|}\right)^2} = 0.0118 \quad (2.34)$$

The fractional error in the rate is 1.18% so the absolute error is $(0.0118)(1.53 \times 10^{-5}) = 1.81 \times 10^{-7}$. This means that the rate is $1.53(0.0181) \times 10^{-5}$ mol/sec.

Multiplication by a constant with no error: A common laboratory activity is to prepare a solution by weighing a mass of solid and then dissolving it in water. The concentration is usually reported in terms of the number of moles of solid rather than the mass. If the mass of NaCl ($W_M = 58.44$) is 40.00(0.01) g, the number of moles is 40.00/58.44 = (40.00)(0.0171) = 0.6845 moles. We can assume that there is no random error associated with the molecular weight of NaCl, so the error for this value is found using Eq. (2.21).

$$\delta q = |0.0171|(0.01) = 0.000171 \quad (2.35)$$

This conversion shows that the amount of NaCl added to the solution is 0.6845(0.00017) mol.

Antilog transformation: The intercept of a double logarithm plot was used to find that log k = −4.18 ± 0.27. Applying the antilog transform shows that k = 6.61 × 10^{-5}. The error for this value is found using Eq. (2.25).

$$\delta k = (2.303)(6.61 \times 10^{-5})(0.27) = 4.11 \times 10^{-5} \quad (2.36)$$

The antilog transformation shows that k = 6.61(4.11) × 10^{-5}.

Logarithmic transformation: If a rate constant is determined to be k = 6.61(4.11) × 10^{-5} the fractional error in k is 0.622. Equation (2.27) is used to find the error in log k.

$$\delta(\log k) = (0.434)(0.622) = 0.270 \quad (2.37)$$

The logarithmic transform of this value shows that log k = −4.18(0.27).

Regression models

Fitting data to equations is an important modeling activity that is supported by an abundance of computer software. The mathematical basis for this software is described in many statistics books and will not be elaborated here. Methodologies that are especially suited for kinetics data are reviewed in Chapter 7 of van Boekel (2009). This section explains some strategies that can guide fitting kinetic data to equations.

One reason to fit data to a function is to summarize the data and report it in a way that allows easy interpolation or other mathematical manipulation. If this is the main objective, the data can be fit to an arbitrary function so long as that function makes a smooth and parsimonious summary of the data and is compatible with desired mathematical manipulations. Typically a low-order polynomial is suitable for this purpose. It is important to avoid over-fitting the data. The goal is to filter out random error without losing the information contained in the data. Increasing the number of regression variables tends to fit the data better as reflected by smaller residuals and a higher correlation coefficient. This is because the additional variables make the fitting function more flexible, with more minima and maxima, so it can pass closer to the data points. However, this additional flexibility can cause the function to flex wildly and to predict values that are beyond the range of the data. This unreasonable result can be avoided by beginning the fitting process using the simplest feasible equation and then adding variables one at a time until the residuals of the fit are about the same size as the errors in the data. Sometimes data can be converted to a linear or other simple form using a logarithmic or power transformation. The fitted equation should not be extrapolated beyond the domain of the original data.

Example 2.4. Finding the rate of Fe(III) reaction with pyrite

Table 2.4 lists Fe^{3+} concentration versus contact time with pyrite from a plug flow reactor experiment (PFR10) (Rimstidt and Newcomb, 1993). The rate of ferric iron consumption is the time derivative of these data. Finding this derivative for $t = 0$ is especially useful because the initial conditions of the experiment are accurately known so their effect on the rate can be clearly established. One strategy for finding the initial rate is to fit the first few concentration versus time

Table 2.4. *Concentration of Fe^{3+} versus contact time between solution and pyrite in a plug flow reactor experiment (PFR10) (Rimstidt and Newcomb, 1993)*

time, sec	Fe^{3+}, molal
0	1.80×10^{-3}
4.7	1.16×10^{-3}
6.3	1.12×10^{-3}
9.1	1.04×10^{-3}
20.5	8.24×10^{-4}

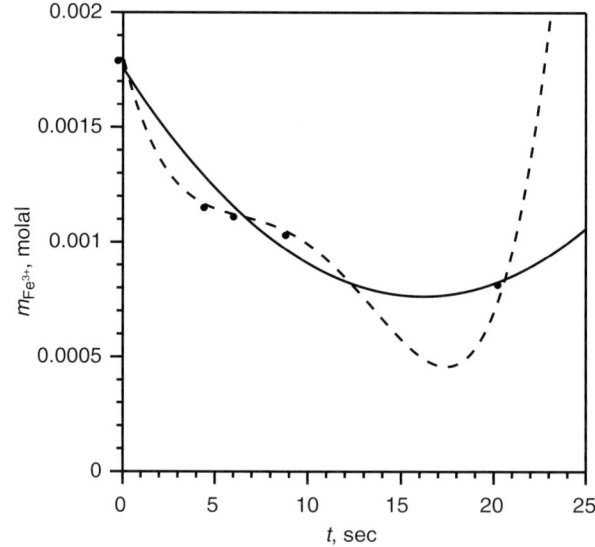

Figure 2.1. Fit of the PFR10 data to a second-order polynomial (solid line, $R^2 = 0.96$) and a fourth-order polynomial (dashed line, $R^2 = 1.00$).

data to a polynomial. Computing the derivative of the polynomial at $t = 0$ gives the initial rate.

$$m_{Fe^{3+}} = a + bt + ct^2 + \cdots \qquad (2.38)$$

$$\frac{dm_{Fe^{3+}}}{dt} = b + 2ct + \cdots \qquad (2.39)$$

When $t = 0$, the rate equals b. Figure 2.1 shows fits of a second and a fourth-order polynomial to the data.

$$m_{Fe^{3+}} = 1.73 \times 10^{-3} - 1.23 \times 10^{-4} t + 3.80 \times 10^{-6} t^2 \qquad (2.40)$$

$$m_{Fe^{3+}} = 1.80 \times 10^{-3} - 3.07 \times 10^{-4} t + 5.34 \times 10^{-5} t^2 - 4.11 \times 10^{-6} t^3 + 1.04 \times 10^{-7} t^4$$

$$(2.41)$$

The second-order fit predicts an initial rate of -1.23×10^{-4} m/sec and the fourth fit predicts a much faster rate of -4.85×10^{-4} m/sec. Both functions fit the data well and the fourth-order polynomial fits it so well that the function passes through all the points so there are no residuals. However, the rate based on the fourth-order polynomial appears to be too fast because that function does not smooth out the random errors in the data. Both functions incorrectly predict that the concentration of Fe^{3+} increases after ~18 seconds of reaction because they are simply approximations for a short interval of the true concentration versus time function.

A second reason to fit data to a function is to test whether the data are consistent with a model or to use a theoretical function to extrapolate experimental results to conditions otherwise not attainable. The goodness of fit of the data to a theoretical function can be gauged by the coefficient of determination (R^2), which is the correlation coefficient squared. R^2 is often interpreted as the fraction of the variability of the response variable that is explained by its functional relationship to the independent variable. For example, if $R^2 = 0.8$ then 80% of the variation in the response variable is explained by the model and 20% of the variation is the result of factors, including random error, that are not part of the model. Some geochemical processes occur under conditions or over time intervals that are difficult to simulate by experiments, making it necessary to use a theoretical model to predict their behavior. This approach uses data from experiments performed under easily attainable conditions to quantify parameters that are used to calibrate a theoretical model. These parameters are then used to predict the value of the variable at

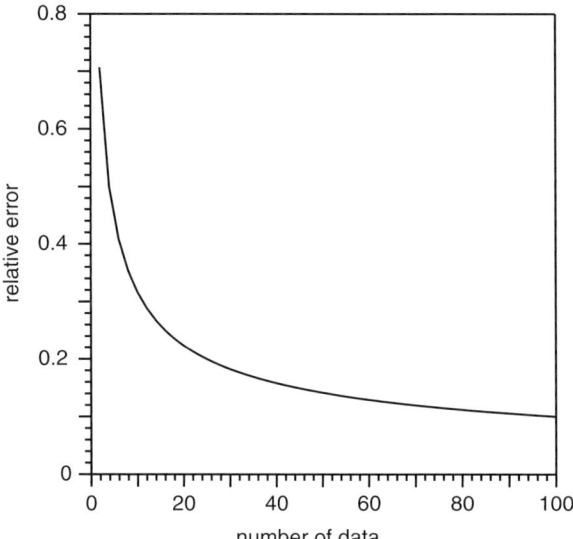

Figure 2.2. The relative error in a statistically estimated value decreases as the reciprocal of the square root of the number of data.

the unattainable conditions. This kind of extrapolation exposes the prediction to two kinds of error. First, the theoretical function may be incorrect or improperly chosen. This possibility is tested by determining (R^2) how well the experimental data fit to the function for the attainable conditions and by comparing predicted values to independently estimated values. The second kind of error is the result of random error in the data that is transformed into uncertainty in the fitted parameters and this uncertainty is further transmitted to the predicted variable. This random error can be significantly reduced by increasing the size of the data set (Gauch Jr, 1993). Figure 2.2 shows that the relative error drops rapidly as the number of data increases. However, this graph also shows that there are diminishing returns for adding data beyond those needed to reduce the error to an acceptable level. Once the data set is established, the regression parameters should be determined using a computer code that calculates their uncertainty. This uncertainty is then forward propagated to provide an estimate of the uncertainty in values predicted by the theoretical function. This approach allows for a quantitative determination of the uncertainty associated with a model's predictions.

Example 2.5. Kaolinite to illite transformation under diagenetic conditions

Rates of many mineral transformations under diagenetic conditions are too slow to measure in the laboratory. In order to model the rate at which kaolinite would transform to illite during burial diagenesis, Chermak and Rimstidt (1990) measured the reaction rates from 250 to 307°C and used the Arrhenius equation

Table 2.5. *Rate constants for the kaolinite to illite conversion reaction*

T, °C	k, sec^{-1}
250	1.18×10^{-10}
275	7.78×10^{-10}
289	1.00×10^{-9}
307	4.48×10^{-9}

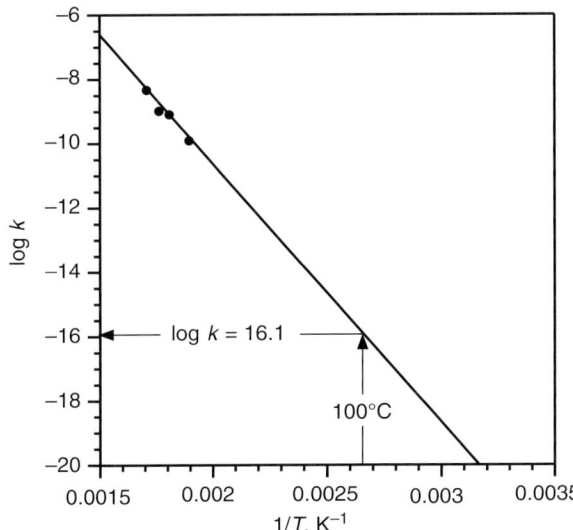

Figure 2.3. Arrhenius plot used to extrapolate rate constants for the kaolinite to illite transformation reaction from high temperature measurements to find the rate constant at 100°C.

to extrapolate those rates to lower temperatures. In this example the goal is to predict the rate constant at 100°C. The rate constants from the high temperature experiments are listed in Table 2.5. These rate constants can be extrapolated to lower temperatures using the Arrhenius equation.

$$k = Ae^{-E_a/RT} \qquad (2.42)$$

This equation can be log-transformed to a linear form.

$$\log k = \log A + \left(\frac{-E_a}{R}\right)\left(\frac{1}{T}\right) \qquad (2.43)$$

Fitting the data in Table 2.5 to this equation gives the line shown in Figure 2.3. This fit explains 97% of the variation in log k ($R^2 = 0.97$).

$$\log k = \frac{-8039(992)}{T} + 5.45(1.8) \quad (2.44)$$

The numbers in parentheses are 1 standard error of the parameter.

Solving this equation for 100°C (373 K) predicts that log k is −16.1 ($k = 7.94 \times 10^{-17}$ sec^{-1}). The uncertainty in the log k value is found by forward propagation of the standard errors associated with the fitting parameters in Eq. (2.44).

$$\delta_{\log k} = \sqrt{(0.18)^2 + (992/T)^2} = 2.67 \quad (2.45)$$

So log k = −16.1(2.7). Even though the data fit the equation quite well, the uncertainty in log k is nearly 3 log units. Most of this error is due to the uncertainty associated with the slope of the line. This example stands as a warning to model builders who extrapolate data without considering error propagation.

Numerical differentiation

Kinetic processes cause a change in a quantity over time. If this quantity is measured at evenly spaced time intervals, it is relatively easy to find the rate by computing a numerical derivative (Pollard, 1977). The simplest method of numerical differentiation is based on the observation that the chord of a graph has a slope that closely approximates the tangent at an intermediate point.

This leads to a simple scheme for calculating the numerical derivative from a data set (Pollard, 1977). The derivative at the first point in the data set (f'_{-1}) gives more weight to the first two points.

$$f'_{-1} = \frac{1}{2h}(-3f_{-1} + 4f_0 - f_1) \quad (2.46)$$

The derivative at each intermediate point (f'_0) is the slope of the chord connecting the surrounding points (Figure 2.4).

$$f'_0 = \frac{1}{2h}(-f_{-1} + f_1) \quad (2.47)$$

The derivative at the last point in the data set (f'_{+1}) gives more weight to the last two points.

$$f'_{+1} = \frac{1}{2h}(f_{-1} - 4f_0 + 3f_1) \quad (2.48)$$

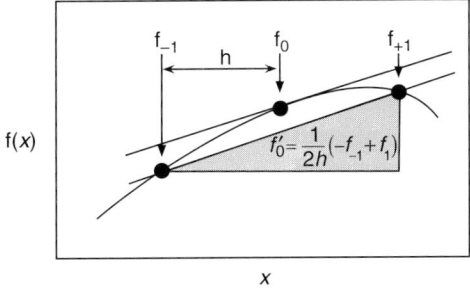

Figure 2.4. Schematic illustration of how the slope of a chord approximates the tangent of a graph at an intermediate point.

Note that the equations for the endpoints of the data set are less dependable than the equation for the intermediate points and they tend to give unreasonable results if the slope is steep. Because numerical differentiation takes the difference between values of the dependent variable it magnifies the scatter in those data. Some of this scatter can be eliminated by various smoothing schemes, but they tend to bias the data and are best avoided.

Example 2.6. Rate of reaction of Fe^{3+} with pyrite in a batch reactor

Ferric iron reacts with pyrite to produce ferrous iron and sulfuric acid.

$$14\ Fe^{3+} + FeS_2(py) + 4\ H_2O = 15\ Fe^{2+} + 2\ SO_4^{2-} + 8\ H^+ \tag{2.49}$$

Rimstidt and Newcomb (1993) measured the concentration of Fe^{3+} in a solution in contact with pyrite every 300 sec ($= h$) for 3000 sec. Their data for experiment BR5 are shown in Table 2.6 and Figure 2.5 along with the numerical derivative, which is the rate of reaction. The computed rates are negative because Fe^{3+} is consumed by the reaction. The rate for the first point ($t = 0$) is found using Eq. (2.46).

$$f'_{-1} = \left(\frac{1}{(2)(300)}\right)\left(-3(9.79 \times 10^{-4}) + 4(9.61 \times 10^{-4}) - 9.45 \times 10^{-4}\right) = -6.33 \times 10^{-8}$$

$$\tag{2.50}$$

The rate for the next point ($t = 300$) is found using Eq. (2.47).

$$f'_0 = \left(\frac{1}{(2)(300)}\right)\left(-9.79 \times 10^{-4} + 9.45 \times 10^{-4}\right) = -5.67 \times 10^{-8} \tag{2.51}$$

The subsequent points are treated the same way. The rate for the last point ($t = 3000$) is found using Eq. (2.48).

$$f'_{+1} = \left(\frac{1}{(2)(300)}\right)\left(9.04 \times 10^{-4} - 4(8.98 \times 10^{-4}) + 3(8.91 \times 10^{-4})\right) = -2.50 \times 10^{-8}$$

$$\tag{2.52}$$

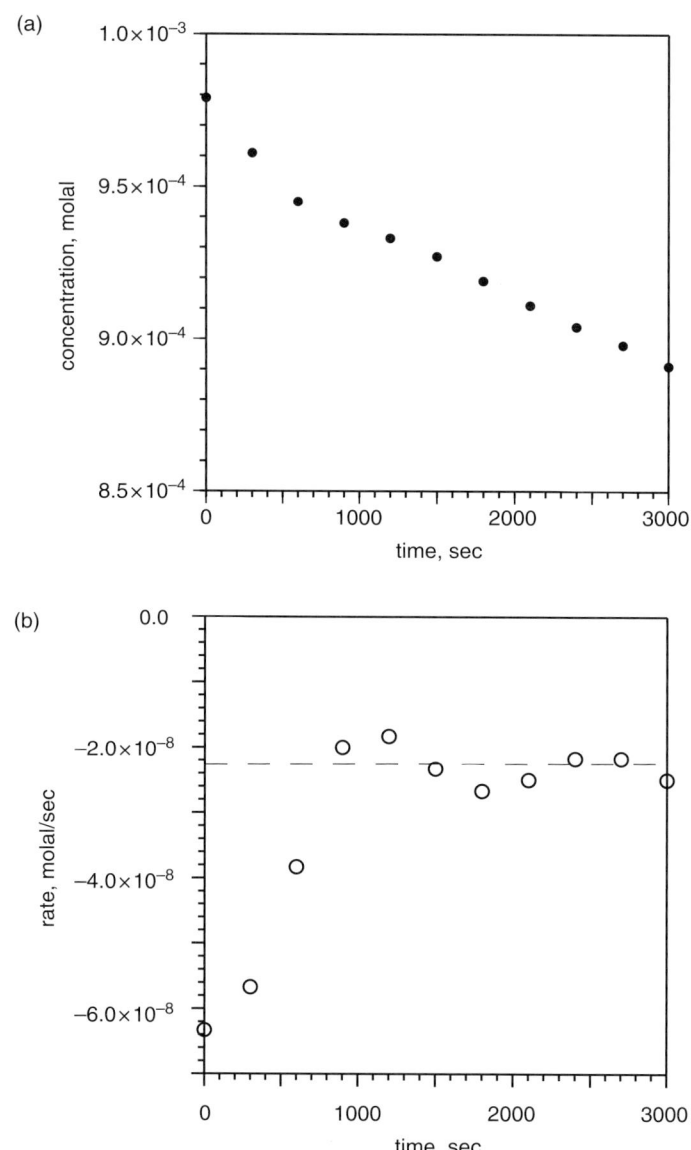

Figure 2.5. (a) Concentration of Fe^{3+} versus time from experiment BR5 (Rimstidt and Newcomb, 1993). (b) Numerical derivatives of the concentration of Fe^{3+} versus time data. Early in the experiment the rates are fast but they slow to a nearly constant rate after ~300 sec.

Table 2.6. *Numerical differentiation of data from experiment BR5 reported in Rimstidt and Newcomb (1993)*

t, sec	m, molal	dm/dt, molal/sec
0	9.79×10^{-4}	-6.33×10^{-8}
300	9.61×10^{-4}	-5.67×10^{-8}
600	9.45×10^{-4}	-3.83×10^{-8}
900	9.38×10^{-4}	-2.00×10^{-8}
1200	9.33×10^{-4}	-1.83×10^{-8}
1500	9.27×10^{-4}	-2.33×10^{-8}
1800	9.19×10^{-4}	-2.67×10^{-8}
2100	9.11×10^{-4}	-2.50×10^{-8}
2400	9.04×10^{-4}	-2.17×10^{-8}
2700	8.98×10^{-4}	-2.17×10^{-8}
3000	8.91×10^{-4}	-2.50×10^{-8}

Chapter 3
Rate equations

Chemical thermodynamics explains why reactions occur and chemical kinetics investigates how they occur. Reacting systems do not reach equilibrium instantly and metastable phases and species can persist for geologic time periods. For example, even though graphite is the thermodynamically stable form of elemental carbon at Earth surface conditions, diamonds have persisted at the Earth's surface for hundreds of millions of years. Metastable persistence can have a profound effect on element distributions and geological processes, so a goal of geochemical kinetics is to model metastable persistence. The time needed for metastable species to convert to stable species is quantified using rate equations. This chapter deals with examples of rate equations that are frequently encountered in geochemical models and shows how the differential rate equations are integrated to quantify metastable persistence.

A chemical reaction converts one or more reactants into one or more products and the rate equation describes how fast this conversion occurs. The reaction rate (r, mol/sec) is the time derivative of the number of moles of chemical species consumed or formed.

$$r = \frac{dn}{dt} = k \prod a_i^{n_i} \tag{3.1}$$

The rate constant, k, incorporates the effects of variables such as temperature, pressure, or ionic strength on the rate. The rate constant can be thought of as a measure of the resistance to the conversion and small values of k produce small values of r. The effect of activity, concentration, or partial pressure of the reacting species on the rate appears in the rate equation as explicit terms each raised to an appropriate power. The product of these terms is the driving force for the reaction, so that large values of activity, partial pressure, or concentration produce large values of r. The exponents for the concentration, activity, or partial pressure terms are called partial orders (n_i) and the sum of those exponents for a rate equation is the overall reaction order, n_o.

$$n_o = \sum n_i \tag{3.2}$$

An elementary reaction step is a reaction that converts reactants directly to products through a single transition state (see Chapter 5). The reaction order for an elementary reaction step usually reflects the molecularity of the reaction. The molecularity of an elementary reaction step is the number of species that come together to form the activated complex.

Composite reactions consist of multiple elementary reaction steps that occur in series, in parallel, or both. Many geochemical reactions are composites of several elementary reaction steps. This makes elucidating their reaction mechanisms very challenging because their reaction order and molecularity are not related in a simple way. Marin and Yablonsky (2011) offer extensive guidance about dealing with composite reactions.

There are many ways to express reaction rates, and keeping track of notation for different kinds of rates along with the units of their accompanying rate equations is challenging. For a simple rate equation such as (3.1), the rate and the rate constant have units of mol/sec, which are the units expected from transition-state theory (Chapter 5). A reaction rate can also be expressed in terms of the time rate of change of concentration of a species (R, mol/kg sec = molal/sec), by dividing both sides of Eq. (3.1) by the mass of water (M) in the system.

$$R = \frac{dm}{dt} = \frac{r}{M} = \frac{1}{M} k \prod a_i^{n_i} \qquad (3.3)$$

If the reaction takes place on an interface that separates two phases and the rate equation expresses how fast a component is transferred to or from that interface, this quantity is actually a flux (J, mol/m²sec).

$$J = \frac{r}{A} = \frac{1}{A} k \prod a_i^{n_i} \qquad (3.4)$$

From Eq. (3.3) we see that $r = RM$ and from Eq. (3.4) we see that $r = JA$. If we combine these expressions we find how the rate of change of concentration in a solution is related to the dissolution flux.

$$R = \frac{dm}{dt} = \frac{A}{M} J = \frac{A}{M} k \prod a_i^{n_i} \qquad (3.5)$$

Equations (3.3) and (3.5) contain both activity and concentration variables. In order to integrate the equation and to make the equations compatible with transition-state theory (Chapter 5), the concentration in the derivative term of these equations can be converted to activity.

$$a = \frac{\gamma \, m}{\gamma^\circ m^\circ} \qquad (3.6)$$

The conventional hypothetical ideal 1 molal standard state standard makes $\gamma° = 1$ and $m° = 1$ mol/kg, so the units on both sides of the equation are sec^{-1}. The resulting equation is simplified by combining all the extraneous terms into an apparent rate constant (k', sec^{-1}).

$$\frac{da}{dt} = \left(\frac{1}{m° M}\frac{\gamma}{\gamma°}\right)r = \left(\frac{\gamma k}{M}\right)\prod a_i^{n_i} = k'\prod a_i^{n_i} \qquad (3.7)$$

This approach can also be applied to Eq. (3.5).

$$\frac{da}{dt} = \left(\frac{A}{m° M}\frac{\gamma}{\gamma°}\right)J = \left(\frac{A}{M}\right)\gamma k \prod a_i^{n_i} = k'\prod a_i^{n_i} \qquad (3.8)$$

More often, the activity terms on the right-hand side are converted to concentration terms by rearranging Eq. (3.6). This allows the equation to be integrated to give the concentration as a function of time, but can result in complicated and often meaningless units for the apparent rate constant. For example, if one or more of the reaction orders is a fraction, as often is the case for mineral dissolution rates, the apparent rate constant will have units of molal raised to a fractional power, which has no physical meaning.

$$R = \frac{dm}{dt} = \left(\frac{k}{M}\right)\prod \gamma_i^{n_i} m_i^{n_i} = \left(\frac{k\prod \gamma_i^{n_i}}{M}\right)\prod m_i^{n_i} = k'\prod m_i^{n_i} \qquad (3.9)$$

$$R = \frac{dm}{dt} = \left(\frac{A}{M}\right)k\prod \gamma_i^{n_i} m_i^{n_i} = \left(\left(\frac{A}{M}\right)k\prod \gamma_i^{n_i}\right)\prod m_i^{n_i} = k'\prod m_i^{n_i} \qquad (3.10)$$

These rate equations describe the rates of production or consumption of chemical species. It is often desirable to report the overall advancement for a chemical reaction in terms of the extent of reaction, ξ (mol). The extent of reaction is defined in terms of the number of moles of species i (n_i, mol) consumed or produced and the stoichiometric coefficient (v_i) for that species.

$$(n_i)_t = (n_i)_{t=0} + v_i \xi \qquad (3.11)$$

The reaction rate is then expressed as the time derivative of the extent of reaction.

$$\frac{d\xi}{dt} = \frac{1}{v_i}\frac{dn_i}{dt} \qquad (3.12)$$

> **Example 3.1.** Oxidation of pyrite by ferric iron
>
> The oxidation of pyrite by ferric iron produces ferrous sulfate and sulfuric acid.
>
> $$14 \text{ Fe}^{3+} + \text{FeS}_2(py) + 8 \text{ H}_2\text{O} = 15 \text{ Fe}^{2+} + 2 \text{ SO}_4^{2-} + 16 \text{ H}^+ \quad (3.13)$$
>
> The reaction rate can be written in terms of any of the species in Eq. (3.13) where the negative sign indicates that the species is consumed.
>
> $$\frac{d\xi}{dt} = -\frac{1}{14}\frac{dn_{\text{Fe}^{3+}}}{dt} = -\frac{1}{1}\frac{dn_{py}}{dt} = -\frac{1}{8}\frac{dn_{\text{H}_2\text{O}}}{dt} = \frac{1}{15}\frac{dn_{\text{Fe}^{2+}}}{dt} = \frac{1}{2}\frac{dn_{\text{SO}_4^{2-}}}{dt} = \frac{1}{16}\frac{dn_{\text{H}^+}}{dt} \quad (3.14)$$
>
> The choice of a reaction progress variable for experimental determination of the reaction rate is limited by practical considerations. Monitoring the rate of water consumption would be nearly impossible because the amount of water consumed is minuscule compared to the 55.5 moles of water per kilogram of solution. Likewise keeping track of the rate of pyrite destruction using a weight-loss technique is very insensitive compared to determining the concentration of an aqueous species. Using the rate of hydrogen ion production as a reaction progress variable is confounded by the fact that the hydrolysis of Fe^{3+} and Fe^{2+} involves hydrogen ions. This leaves measuring the concentration of Fe^{3+}, Fe^{2+}, or SO_4^{2-} concentrations as the best way for monitoring the reaction progress. Ideally the rate of consumption of Fe^{3+} along with the rates of production of Fe^{2+} and SO_4^{2-} would be determined and compared to see if they are consistent with Eq. (3.14). For example, if the rate of production of sulfate is not ~1/7 the rate of Fe^{3+} consumption, either the analytical procedure has a systematic error or reactions in addition to Eq. (3.13) are consuming or producing these species.

Integrated rate equations

Many geochemical models are based on the integrated form of rate equations. Some typical examples are shown here and the integrated versions of other forms can be found in various kinetics textbooks. Capellos and Beielski (1972) give a comprehensive treatment of integrated rate equations for simple as well as some composite reactions.

Unopposed reactions (n ≠ 1)

When the rate for the reverse reaction is sufficiently small it can be ignored, so the overall reaction rate equals the rate of the forward reaction.

$$A \xrightarrow{R_+} \text{products} \quad (3.15)$$

In the simplest case, only one aqueous species affects the reaction rate and the rate of consumption of species A is proportional to the rate constant (k) multiplied by its concentration (m, mol/kg) raised to an arbitrary power (n). Values of n and k are determined empirically.

$$\frac{dm}{dt} = -km^n \tag{3.16}$$

This equation is rearranged and integrated between the initial ($t = 0$, $m = m_o$) and final conditions ($t = t$, $m = m$) to produce a relationship between the concentration of A and time.

$$\int_{m_o}^{m} \frac{dm}{m^n} = -k \int_0^t dt \tag{3.17}$$

$$m^{(-n+1)} - m_o^{(-n+1)} = -(-n+1)kt \tag{3.18}$$

$$\left(\frac{m}{m_o}\right)^{(-n+1)} = 1 - \frac{(-n+1)kt}{m_o^{(-n+1)}} \tag{3.19}$$

Equation (3.19) can be simplified by recasting it in terms of the fraction of material remaining (p).

$$p = \frac{m}{m_o} \tag{3.20}$$

$$p = \left(1 - \frac{(-n+1)kt}{m_o^{(-n+1)}}\right)^{(1/(-n+1))} \tag{3.21}$$

In some cases it is better to recast the equation in terms of the fraction reacted ($\alpha = 1 - p$).

$$\alpha = 1 - \left(1 - \frac{(-n+1)kt}{m_o^{(-n+1)}}\right)^{(1/(-n+1))} \tag{3.22}$$

Example 3.2. Dissolved oxygen consumption by reaction with pyrite

When air-saturated surface waters infiltrate pyritic mine wastes, the dissolved oxygen (DO) is consumed by reaction with pyrite.

$$FeS_2 + 7/2\ O_2 + H_2O = Fe^{2+} + 2\ H^+ + 2\ SO_4^{2-} \tag{3.23}$$

Equation (3.21) can be used to find how long it will take for all the DO to be consumed if 1 kg of air-saturated solution ($m_{DO} = 2.7 \times 10^{-4} = 10^{-3.57}$ mol/kg) contacts 1 m² of pyrite surface.

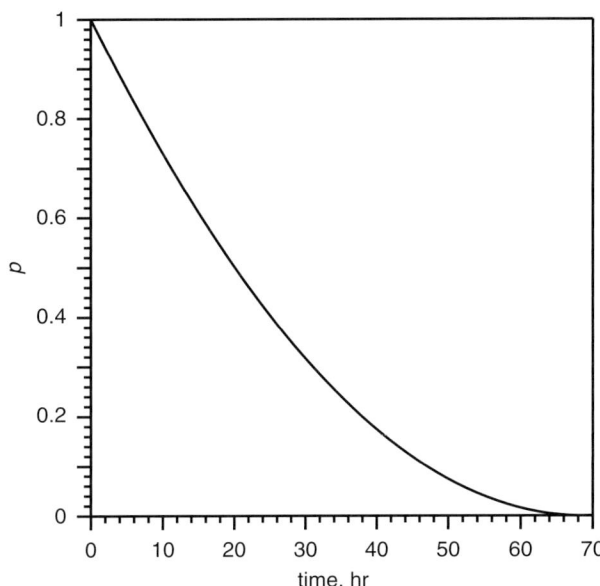

Figure 3.1. Fraction of DO remaining as a function of time for 1 kg of air-saturated solution reacting with 1 m² of pyrite surface.

Williamson and Rimstidt (1994) report the following rate equation for reaction (3.23).

$$J_{py} = A\left(10^{-8.19} m_{DO}^{0.5} m_{H^+}^{-0.11}\right) \quad (3.24)$$

The rate of DO consumption is related to the rate of pyrite destruction by the stoichiometry of Eq. (3.23).

$$\frac{d\xi}{dt} = -J_{py} = -\frac{2}{7}J_{DO} \quad (3.25)$$

Equations (3.5), (3.24), and (3.25) can be combined to produce a rate equation for DO consumption for a system with 1 kg of water and 1 m² of pyrite surface area.

$$R_{DO} = -\left(\frac{7}{2}\right)\left(10^{-8.19} 10^{0.77}\right) m_{DO}^{0.5} = -\left(10^{0.544} 10^{-8.19} 10^{0.77}\right) m_{DO}^{0.5} = -10^{-6.88} m_{DO}^{0.5} \quad (3.26)$$

In this equation, the rate constant, $k = 10^{-6.88}$; the initial DO concentration, $m_o = 10^{-3.56}$; and the reaction order, $n = 0.5$. If α is set equal to zero in Eq. (3.21), the resulting relationship can be rearranged to give the time needed to completely consume the DO.

$$t = \frac{m_o^{(1-n)}}{(1-n)k} = \frac{\left(10^{-3.57}\right)^{0.5}}{10^{-0.30}10^{-6.88}} = 10^{5.40} = 2.48 \times 10^5 \text{ sec} = 69.0 \text{ hr} \qquad (3.27)$$

This can be verified by using Eq. (3.21) to create a graph (Figure 3.1) of p as a function of time.

Unopposed reactions (n = 1)

The above model applies for all values of n other than one. However first-order reactions ($n = 1$) are relatively common and the integrated form of a first-order unopposed rate equation is much simpler.

$$\left(\frac{dm}{dt}\right) = -km \qquad (3.28)$$

Rearranging and integrating this equation between the initial ($t = 0$, $m = m_o$) and final conditions ($t = t$, $m = m$) gives an equation for the concentration as a function of time.

$$\int_{m_o}^{m} \frac{dm}{m} = -k \int_0^t t \qquad (3.29)$$

$$\ln\left(\frac{m}{m_o}\right) = -kt \qquad (3.30)$$

It is often useful to express this result in terms of the fraction reacted (α) or the fraction remaining (p).

$$\alpha = 1 - p = \frac{m}{m_o} = e^{-kt} \qquad (3.31)$$

Characteristic time

Reaction rates are often compared using a characteristic time. For example, the characteristic time might be defined as the time needed for the destruction of 50% of the reactants ($\alpha = p = 0.5$). This time is called the reaction's half-life, $t_{1/2}$. When $t = t_{1/2}$, $\alpha = p = 0.5$. The half-life for an unopposed reaction where $n \neq 1$ is found by setting $m = 0.5$ in Eq. (3.18).

$$t_{1/2} = \frac{m_o^{(-n+1)} - (0.5m_o)^{(-n+1)}}{(-n+1)k} \qquad (3.32)$$

The half-life for an unopposed reaction where $n = 1$ is found by setting $m = 0.5$ in Eq. (3.30).

$$t_{1/2} = \frac{\ln 0.5}{-k} = \frac{0.693}{k} \tag{3.33}$$

The time constant is another characteristic time that is sometimes used to describe processes that follow exponential growth or decay as described by Eq. (3.34). The time constant is defined as $t_c = 1/k$ (sec). If $t = t_c$, the reaction has proceeded to 63.2% of completion.

$$p = 1 - e^{-1} = 1.0 - 0.368 = 0.632 \tag{3.34}$$

After $5\,t_c$, $p = 99.3$ and for all practical purposes the reaction can be said to be complete.

Example 3.3. Oxidation of hydrogen sulfide

Hydrogen sulfide is produced as a byproduct of the decay of organic matter in sediments. When this H_2S escapes to the overlying water column, it is oxidized by reaction with dissolved oxygen. The persistence of this hydrogen sulfide is strongly dependent upon pH, so it is instructive to express the half-life of H_2S in air-saturated fresh water ($I \approx 0$) as a function of pH.

The reaction is $H_2S + O_2 \rightarrow$ products. This reaction involves several electron transfer steps with one, or at most two, electrons transferred per elementary step, so it is not surprising that there are several observed reaction intermediates including SO_3^{2-}, SO_4^{2-}, and $S_2O_3^{2-}$ (Avrahami and Golding, 1968; Zhang and Millero, 1993).

In spite of the complexity of the reaction, the rate equation that describes H_2S disappearance is first order in H_2S and first order in O_2, making it an overall second-order reaction (Millero, 2001).

$$R_{H_2S} = -k m_{H_2S} m_{O_2} \tag{3.35}$$

This is an important and well-studied reaction so the effects of temperature and ionic strength on k are known. When they are taken into consideration, the rate constant (k, hr^{-1}) can be expressed in terms of the hydrogen ion concentration.

$$k = \frac{14.67 + 2.84 \times 10^{-7} m_{H^+}^{-1.16}}{1 + 1.05 \times 10^{-7} m_{H^+}^{-1.0}} \tag{3.36}$$

The concentration of dissolved oxygen in equilibrium with air is 2.7×10^{-4} molal. This can be combined with the expression for the rate constant above to find an overall rate constant, k'.

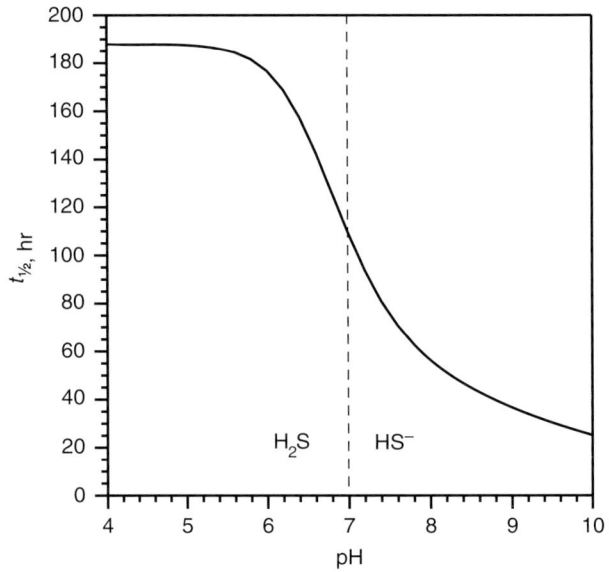

Figure 3.2 Half-life of H_2S in air-saturated solution as a function of pH at 25°C. The rapid decrease in half-life near pK_1 for H_2S suggests that HS^- is much more reactive than H_2S.

$$k' = \left(2.70 \times 10^{-4}\right)\left(\frac{14.67 + 2.85 \times 10^{-7} m_{H^+}^{-1.16}}{1 + 1.05 \times 10^{-7} m_{H^+}^{-1.0}}\right) \quad (3.37)$$

Combining Eqs (3.35) and (3.37) gives an equation that expresses the rate as a function of hydrogen sulfide concentration.

$$R_{H_2S} = \frac{dm_{H_2S}}{dt} = \left(\frac{3.96 \times 10^{-3} + 6.70 \times 10^{-11} m_{H^+}^{-1.16}}{1 + 1.05 \times 10^{-7} m_{H^+}^{-1.0}}\right) m_{H_2S} = k' m_{H_2S} \quad (3.38)$$

This equation is integrated to produce an expression for the fraction remaining as a function of time.

$$\ln\left(\frac{m}{m_o}\right) = \ln p = -k't \quad (3.39)$$

$$p = e^{-(k't)} \quad (3.40)$$

Substituting $\alpha = 0.5$ into this equation gives a relationship for the half-life of H_2S ($t_{1/2}$, hr) as a function of hydrogen ion concentration.

$$t_{1/2} = \frac{\ln 0.5}{-k'} = \frac{0.693}{k'} = \frac{0.693}{\left(\dfrac{3.96 \times 10^{-3} + 6.70 \times 10^{-11} m_{H^+}^{-1.16}}{1 + 1.05 \times 10^{-7} m_{H^+}^{-1.0}}\right)} \quad (3.41)$$

$$t_{1/2} = \frac{0.693(1 + 1.05 \times 10^{-7} m_{H^+}^{-1.0})}{3.96 \times 10^{-3} + 6.70 \times 10^{-11} m_{H^+}^{-1.16}} = \frac{0.693 + 6.71 \times 10^{-8} m_{H^+}^{-1.0}}{3.96 \times 10^{-3} + 6.70 \times 10^{-11} m_{H^+}^{-1.16}} \text{ hr} \quad (3.42)$$

A graph of $t_{1/2}$ versus pH (Figure 3.2) shows that the half-life of H_2S is relatively long and constant between pH 4 to 6, but declines rapidly as the pH approaches pK_1 of H_2S (6.98), suggesting that HS^- oxidizes more quickly than H_2S.

Opposed reactions: principle of detailed balance

If a reaction proceeds in both the forward and backward direction simultaneously, the concentrations of the reacting species change until the reaction reaches equilibrium.

$$A \underset{R_-}{\overset{R_+}{\rightleftarrows}} B \quad (3.43)$$

For the very simple reaction described by Eq. (3.43), the rate of the forward reaction is the product of the forward rate constant (k_+) and the activity of A.

$$R_- = \frac{dm_B}{dt} = k_+ a_A \quad (3.44)$$

The rate of the backward reaction is the product of the backward rate constant (k_-) and the activity of B.

$$R_+ = \frac{dm_A}{dt} = k_- a_B \quad (3.45)$$

The overall reaction rate is the difference between the forward and backward rates.

$$R = k_+ a_A - k_- a_B \quad (3.46)$$

Regardless of the original activities of A and B, their activities will change until the system reaches equilibrium. The principle of detailed balance (also called microscopic reversibility) states that at equilibrium the rate of the forward reaction is exactly equal to the rate of the backward reaction so that the net rate is zero. When $R_A = R_B$, $k_+ a_{A,\,eq} = k_- a_{B,\,eq}$, and that means that the equilibrium constant is the quotient of the rate constants, $K = k_+/k_-$.

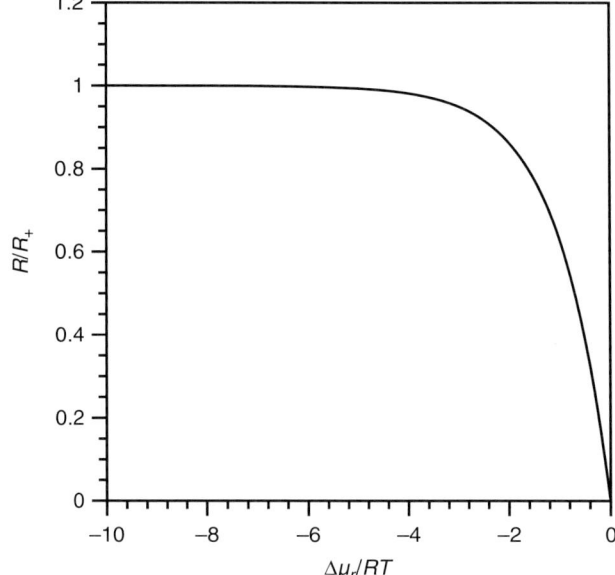

Figure 3.3. Reaction rate normalized to the far-from-equilibrium rate versus chemical potential driving the reaction. The overall rate approaches zero as the driving force for the reaction approaches zero.

$$\frac{k_+}{k_-} = \frac{a_{B,\,eq}}{a_{A,\,eq}} = K \qquad (3.47)$$

This relationship appears to be valid even when the reaction is not at equilibrium so long as the reaction mechanism (i.e. transition state) remains the same. This means that the rate equation can also be written to make the rate proportional to the activity ratio, $Q = a_A/a_B$.

$$R = -k_+ a_A \left(1 - \frac{k_- a_B}{k_+ a_A}\right) = -k_+ a_A \left(1 - \frac{Q}{K}\right) \qquad (3.48)$$

The chemical potential driving the reaction is a function of Q/K.

$$\Delta\mu_r = -RT \ln\left(\frac{Q}{K}\right) \qquad (3.49)$$

$$\frac{Q}{K} = e^{\frac{-\Delta\mu_r}{RT}} \qquad (3.50)$$

Combining Eqs (3.48) and (3.50) gives an equation that relates the reaction rate to the chemical potential driving the reaction. This relationship can also be expressed in terms of affinity where $A = -\Delta\mu_r$. Affinity is a key concept in non-equilibrium thermodynamics, which is discussed briefly in Chapter 10.

$$R = k_+ a_A \left(1 - e^{\frac{-\Delta\mu_r}{RT}}\right) \qquad (3.51)$$

When the reaction is far from equilibrium ($\Delta\mu_r < \sim -3\,RT$) the reverse reaction rate is effectively zero, so the far-from-equilibrium reaction rate (R_+) $\approx k_+ a_A$. This means that Eq. (3.51) can be written to show how the overall rate declines as the reaction approaches equilibrium (Figure 3.3).

$$\frac{R}{R_+} = \frac{R}{k_+ a_A} = 1 - e^{\frac{-\Delta\mu_r}{RT}} \qquad (3.52)$$

This relationship is valid only if the reaction mechanism remains the same over the entire range of conditions. If the reaction mechanism changes, the relationship between R and μ becomes complicated.

Opposed reaction: first-order forward and first-order backward

The simplest form of an opposed reaction involves only one reactant and one product species.

$$A \underset{R_B = k_- m_B}{\overset{R_A = k_+ m_A}{\rightleftarrows}} B \qquad (3.53)$$

The net rate of appearance of B is the difference between the forward and reverse rates.

$$\frac{dm_B}{dt} = R_B - R_A = k_+ m_A - k_- m_B \qquad (3.54)$$

Aris (1989) presents three different schemes to integrate this differential rate equation for an initial condition where the concentration of B is zero, i.e. $m_{Bo} = 0$ when $t = 0$. The following derivation is a more general derivation where $m_{Bo} \geq 0$. The trick to this derivation is to identify relationships, which allows Eq. (3.54) to be recast into a form that is easily integrated. The most useful of these relationships are the conservation of mass and the principle of detailed balance. The conservation of mass requires that the concentration of A and B must sum to a constant value so that the sum of concentrations of A and B at any time must equal the sum of their concentrations at equilibrium ($m_{Ae} + m_{Be}$).

$$m_A + m_B = m_{Ae} + m_{Be} \qquad (3.55)$$

The principle of detailed balance requires that the quotient of the equilibrium concentrations be equal to the equilibrium constant. This applies for dilute solutions where the activity coefficients are very near one.

$$K = \frac{k_+}{k_-} = \frac{m_{Be}}{m_{Ae}} \tag{3.56}$$

The conservation of mass equation (3.55) can be used to write the m_A term in Eq. (3.54) in terms of m_B.

$$\frac{dm_B}{dt} = k_+\left((m_{Ae} + m_{Be}) - m_B\right) - k_- m_B \tag{3.57}$$

Rearranging and integrating this equation between the initial ($t = 0$, $m_B = m_{Bo}$) and final conditions ($t = t$, $m_B = m_B$) gives an equation for the concentration of B as a function of time.

$$\frac{dm_B}{dt} = k_+(m_{Ae} + m_{Be}) - (k_+ + k_-)m_B \tag{3.58}$$

$$\int_{m_{Bo}}^{m_B} \frac{dm_B}{k_+(m_{Ae} + m_{Be}) - (k_+ + k_-)m_B} = \int_0^t dt \tag{3.59}$$

The integral is evaluated between the initial and final conditions.

$$\ln\left[\frac{k_+(m_{Ae} + m_{Be}) - (k_+ + k_-)m_B}{k_+(m_{Ae} + m_{Be}) - (k_+ + k_-)m_{Bo}}\right] = -(k_+ + k_-)t \tag{3.60}$$

The m_A term is eliminated by substitution of a rearranged version of Eq. (3.56); $m_A = m_{Be}/K$.

$$\ln\left[\frac{k_+\left(m_{Be}/K + m_{Be}\right) - (k_+ + k_-)m_B}{k_+\left(m_{Be}/K + m_{Be}\right) - (k_+ + k_-)m_{Bo}}\right] = -(k_+ + k_-)t \tag{3.61}$$

Next the equilibrium constant is eliminated using the principle of detailed balance, $k_+/K = k_-$.

$$\ln\left[\frac{(k_+ + k_-)m_{Be} - (k_+ + k_-)m_B}{(k_+ + k_-)m_{Be} - (k_+ + k_-)m_{Bo}}\right] = \ln\left[\frac{m_{Be} - m_B}{m_{Be} - m_{Bo}}\right] = -(k_+ + k_-)t \tag{3.62}$$

$$\ln\left[\frac{m_{Be} - m_B}{m_{Be} - m_{Bo}}\right] = \ln\left[\frac{1 - \frac{m_B}{m_{Be}}}{1 - \frac{m_{Bo}}{m_{Be}}}\right] = (k_+ + k_-)t \tag{3.63}$$

This equation can be transformed and rearranged to give the ratio of the concentration of B to its equilibrium concentration.

$$\left(\frac{m_B}{m_{Be}}\right) = 1 - \left(1 - \frac{m_{Bo}}{m_{Be}}\right)e^{-(k_+ + k_-)t} \tag{3.64}$$

If $m_{Bo} = 0$, this equation is relatively simple.

$$\left(\frac{m_B}{m_{Be}}\right) = 1 - e^{-(k_+ + k_-)t} \tag{3.65}$$

$$m_B = m_{Be}\left(1 - e^{-(k_+ + k_-)t}\right) \tag{3.66}$$

Equation (3.65) can be rearranged to find a characteristic time, defined as the time needed for m_B to change from 0 to 0.5 m_{Be}.

$$0.5 = \frac{m_B}{m_{Be}} = \left(1 - e^{-(k_+ + k_-)t_{1/2}}\right) \tag{3.67}$$

$$t_{1/2} = \frac{\ln 0.5}{-(k_+ + k_-)} = \frac{-0.693}{-(k_+ + k_-)} = \frac{0.693}{k_+ + k_-} \tag{3.68}$$

Example 3.4. Hydration of CO_2

CO_2 dissolves in water as $CO_2(aq)$, which then reacts with water to form carbonic acid, H_2CO_3.

$$CO_2(aq) + H_2O \underset{R_-}{\overset{R_+}{\rightleftarrows}} H_2CO_3 \tag{3.69}$$

This hydration–dehydration reaction is slow enough to make its rate significant in some biochemical processes and it may be slow enough to influence the course of certain geochemical processes. We can get a sense of how fast this reaction proceeds by calculating its half-life along with the time needed to achieve 99% of equilibrium.

The rate of the forward reaction is the forward rate constant times the concentration of $CO_2(aq)$. Water is in great excess so its activity is set equal to 1.

$$R_+ = \frac{dm_{H_2CO_3}}{dt} = k_+ a_{CO_2} a_{H_2O} = k_+ a_{CO_2} \tag{3.70}$$

The rate of the reverse reaction is the reverse rate constant times the concentration of H_2CO_3.

Table 3.1. *Values of k_+ and k_- at 25°C compiled by Garg and Maren (1972) along with the average of the values. The standard error of the mean is given in parentheses.*

	k_+, molal/sec	k_-, molal/sec
	0.0275	31
	0.0257	25.9
	0.026	20.6
	0.043	25.5
	0.0358	27
	0.0375	15.1
	0.044	17.5
	0.036	13.7
	0.037	18
		16.7
Average	0.0347(0.002)	21.1(1.9)

$$R_- = \frac{dm_{CO_2}}{dt} = k_- a_{H_2CO_3} \qquad (3.71)$$

The reacting species have zero charge, so we can set their concentrations equal to their activity. The overall rate of appearance of H_2CO_3 is the difference between its rate of formation and the rate of destruction as defined by Eqs (3.70) and (3.71).

$$R_+ - R_- = \frac{dm_{H_2CO_3}}{dt} = k_+ m_{CO_2} - k_- m_{H_2CO_3} \qquad (3.72)$$

It is not uncommon to find multiple determinations of rate constants for important reactions like the CO_2 hydration/dehydration reaction. Table 3.1 lists several values of k_+ and k_- that were tabulated in Garg and Maren (1972). Each of these values came from an independent experimental study, so a reasonable way to find the most probable values for these rate constants is to average the reported values.

The principle of detailed balance can be used along with the average values for the rate constants to calculate the equilibrium constant for reaction (3.69).

$$K = \frac{a_{H_2CO_3}}{a_{CO_2}} = \frac{k_+}{k_-} = \frac{0.0347}{21.1} = 1.64 \times 10^{-3} \qquad (3.73)$$

This relatively small value means that at equilibrium only 0.16% of the dissolved CO_2 is in the form of H_2CO_3. This is due to the small size of the forward rate

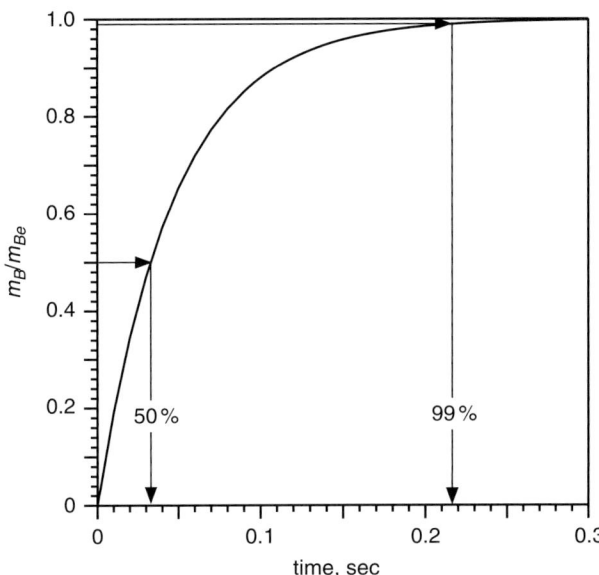

Figure 3.4. Graph of m_B/m_{Be} versus time showing the time needed for the H_2CO_3 concentration reaction to reach 50% and 99% of the equilibrium value.

constant relative to the backward rate constant, i.e. H_2CO_3 forms slowly and dissociates quickly.

Equation (3.66) can be rearranged to find the time needed for H_2CO_3 to reach a fraction of the equilibrium concentration.

$$t = -\frac{1}{(k_+ + k_-)} \ln\left(1 - \frac{m_B}{m_{Be}}\right) \tag{3.74}$$

The graph of m_B/m_{Be} versus time is shown in Figure 3.4.
If $m_B/m_{Be} = 0.5$,

$$t = -\frac{1}{(21.13)} \ln(1 - 0.5) = -(0.0473)(-0.301) = 0.0327 \sec \tag{3.75}$$

If $m_B/m_{Be} = 0.99$,

$$t = -\frac{1}{(21.13)} \ln(1 - 0.99) = -(0.0473)(-4.61) = 0.218 \sec \tag{3.76}$$

Although these times may seem short compared to the characteristic times for many geochemical processes, they are actually very slow compared to the rates of interaction of most aqueous ions with water molecules.

Opposed reaction: zero-order forward and first-order backward

In some cases, the activity (or concentration) of the reactants remains constant over the duration of a reversible reaction. Mineral dissolution, where mineral A dissolves to form aqueous species B, is a good example of this situation.

$$A(s) \underset{R_B = k'_- a_B}{\overset{R_A = k'_+ a_A = k'_+}{\rightleftarrows}} B(aq) \tag{3.77}$$

Phase A is a solid, so $a_A = 1$ at all times. Because the reaction occurs at the interface between the solid and the solution, the rate is directly proportional to the area of that interface. This proportionality is incorporated into apparent forward (k'_+) and reverse (k'_-) rate constants.

$$k'_+ = A k_+ \text{ and } k'_- = A k_- \tag{3.78}$$

The overall rate of appearance (or disappearance) of B is the difference between the dissolution and precipitation rate.

$$R_B - R_A = \frac{da_B}{dt} = k'_+ - k'_- a_B \tag{3.79}$$

This equation can be rearranged, integrated, and evaluated from an initial condition where $a_B = a_{B_o}$ when $t = 0$ to $a_B = a_B$ when $t = t$.

$$\int_{a_{B_o}}^{a_B} \frac{da_B}{k'_+ - k'_- a_B} = \int_0^t dt \tag{3.80}$$

$$\ln\left(\frac{k'_+ - k'_- a_B}{k'_+ - k'_- a_{B_o}}\right) = -k'_- t \tag{3.81}$$

$$\ln\left(\frac{1 - \frac{k'_- a_B}{k'_+}}{1 - \frac{k'_- a_{B_o}}{k'_+}}\right) = \ln\left(\frac{1 - \frac{a_B}{K}}{1 - \frac{a_{B_o}}{K}}\right) = -k'_- t \tag{3.82}$$

The equilibrium constant for Eq. (3.77) is equal to the equilibrium activity of species B ($K = a_{B_e}$) so K in Eq. (3.82) can be replaced by a_{B_e}.

$$\left(\frac{a_B}{a_{B_e}}\right) = 1 - \left(1 - \frac{a_{B_o}}{a_{B_e}}\right) e^{-k'_- t} \tag{3.83}$$

If no B is present in solution at the beginning of the reaction, $a_B / a_{B_e} = 0$. This activity ratio increases as the reaction proceeds, so that when $a_B / a_{B_e} = 1$ the reaction has reached equilibrium.

$$\left(\frac{a_B}{a_{B_e}}\right) = 1 - e^{-k'_- t} \tag{3.84}$$

This equation can be rearranged to find a characteristic time ($t_{1/2}$, sec), which is defined as the time needed for a_B/a_{B_e} to change from 0 to 0.5.

$$0.5 = \frac{a_B}{a_{Be}} = \left(1 - e^{-k'_- t_{1/2}}\right) \tag{3.85}$$

$$t_{1/2} = \frac{\ln 0.5}{-k'_-} = \frac{-0.693}{-k'_-} = \frac{0.693}{k'_-} \tag{3.86}$$

Example 3.5. Quartz equilibrium times

Quartz makes up about 15% of the Earth's crust, so it is not surprising that it often controls the dissolved silica concentration in adjacent pore waters, especially at high temperatures. However, at low temperatures the ability of quartz to buffer the dissolved silica concentration is limited by its slow dissolution and precipitation rates. We can visualize the time needed for a solution to equilibrate with quartz as a function of temperature using the rate constants from Rimstidt and Barnes (1980) to construct a graph showing the concentration (activity) of dissolved silica versus time for solutions that initially contain no dissolved silica at temperatures of 25°, 50°, and 75°C. This model considers a system consisting of 1 m² of quartz in contact with 1 kg of water.

Quartz dissolves to release silicic acid into solution.

$$SiO_2(qz) + 2\ H_2O(l) = H_4SiO_4(aq) \tag{3.87}$$

$$K = a_{H_4SiO_4 e} \tag{3.88}$$

The differential rate equation describing quartz dissolution is analogous to Eq. (3.79).

$$\frac{da_{H_4SiO_4}}{dt} = k'_+ - k'_- a_{H_4SiO_4} \tag{3.89}$$

The integrated form of this rate equation is analogous to Eq. (3.81).

$$\ln\left(\frac{k'_+ - k'_- a_{H_4SiO_4}}{k'_+ - k'_- a_{H_4SiO_4, o}}\right) = -k'_- t \tag{3.90}$$

If there is no dissolved silica when $t = 0$, the integrated rate equation becomes analogous to Eq. (3.84).

$$a_{H_4SiO_4} = \left(a_{H_4SiO_4 e}\right)\left(1 - e^{-k'_- t}\right) \tag{3.91}$$

Table 3.2. *Values of the precipitation rate constant and the equilibrium constant needed for this calculation.*

T, °C	K	k'_-
25	1.11×10^{-4}	3.80×10^{-10}
50	2.50×10^{-4}	1.79×10^{-9}
75	4.94×10^{-4}	6.77×10^{-9}

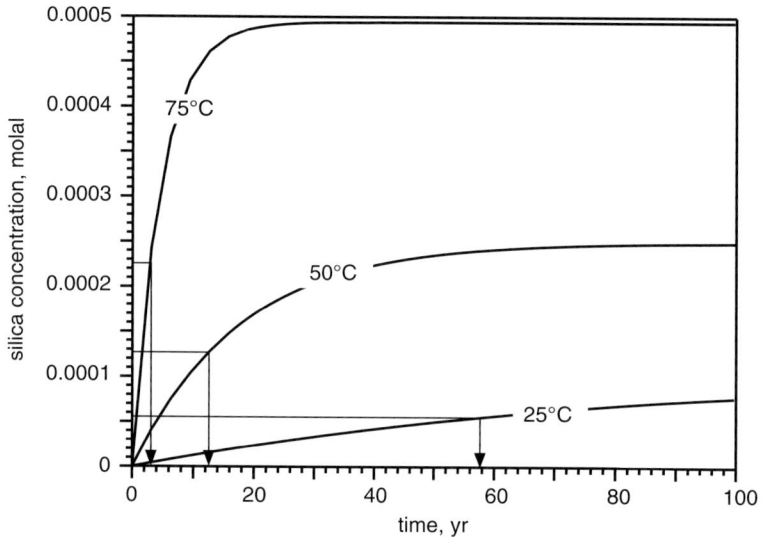

Figure 3.5. Graph showing the concentration versus time values for 1 m² of quartz surface dissolving into 1 kg of water. The vertical lines show the time needed to reach 50% of the equilibrium concentration for each temperature.

Because H_4SiO_4 is an uncharged species, we can assume that its activity coefficient is near 1, and because the activity of quartz and water are 1 we can set $a_{H_4SiO_4 e} = K$.

This allows Eq. (3.91) to be rewritten in terms of the equilibrium constant and the precipitation rate constant.

$$m_{H_4SiO_4} = K\left(1 - e^{-k'_- t}\right) \qquad (3.92)$$

The values of the precipitation rate constant and the equilibrium constant at 25°, 50°, and 75°C (Rimstidt and Barnes, 1980) given in Table 3.2 were used to calculate the concentration versus time graphs shown in Figure 3.5.

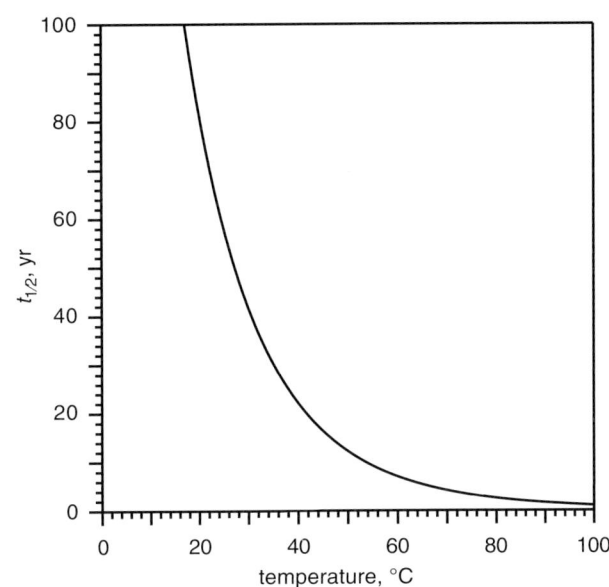

Figure 3.6. $t_{1/2}$ for the quartz precipitation–dissolution reaction as a function of temperature.

Figure 3.5 shows that even though the equilibrium concentration of dissolved silica increases considerably with increasing temperature, the time needed to reach 50% of that equilibrium concentration decreases significantly from 58 years at 25°C to 3.2 years at 75°C.

Another way to visualize this is to create a graph that shows $t_{1/2}$ as a function of temperature using the relationship $t_{1/2} = 0.693/k'_-$ where $\log k_- = -0.707 - 2589/T$ (K) (Rimstidt and Barnes, 1980). This graph is shown in Figure 3.6. At temperatures above 100°C the equilibration rates become relatively fast.

Chapter 4
Chemical reactors

Reactor models define the environment in which chemical reactions occur. Chemical engineers developed the concept of ideal chemical reactors to model processing equipment in chemical plants (Aris, 1956; Hill Jr., 1977; Levenspiel, 1972a, 1972b; Schmidt, 1998). Figure 4.1 summarizes the essential features of the three types of ideal chemical reactors discussed in this chapter. Batch reactors (BR) are well mixed and they are closed to mass transfer but open to energy transfer making them equivalent to closed thermodynamic systems. Mixed flow reactors (MFR) are also well mixed and they are open to both mass and energy transfer so they are equivalent to open thermodynamic systems. Engineers usually refer to mixed flow reactors as continuously stirred tank reactors (CSTR) and geochemists sometimes call them "flow through" reactors. Plug flow reactors (PFR) are also open to both mass and energy transfer but the reacting solution passes through the reactor as slugs that do not mix with each other. The slugs of fluid are closed to fluid loss or gain so they act as if they are batch reactors moving through space.

Quantitative models of reactor performance are based on the principles of conservation of mass and energy. This chapter focuses on the conservation of mass, but modeling the energy budget of reactors uses the same approach. The rate of accumulation of a substance in the reactor equals the rate of addition minus the rate of removal plus the rate of generation.

$$r_{acc} = r_{in} - r_{out} + r_{gen} \tag{4.1}$$

Ideal chemical reactors can be linked in many different ways so that the output of one reactor becomes the input of the next so that the reactions occur in stages. Staged reactor models can be adapted to simulate most, if not all, real-world scenarios. This means that the behavior of complex interacting systems can be simplified to combinations of simple reactors, each of which is relatively easy to model. Ideal chemical reactor models can be applied as a first approximation to most natural situations and the concept easily leads to mathematical descriptions of those situations. For example,

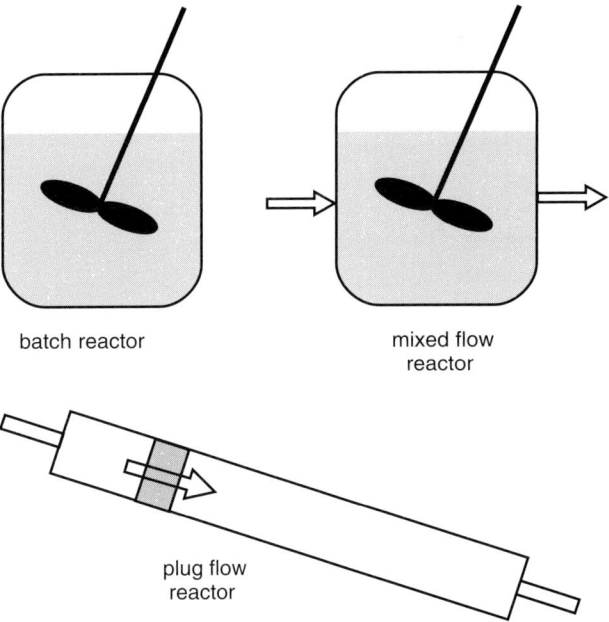

Figure 4.1. The three types of ideal chemical reactors. The ideal batch reactor (BR) is well mixed but closed to mass transfer. The ideal mixed flow reactor (MFR) is well mixed and subject to continuous mass transfer. The fluid in an ideal plug flow reactor (PFR) moves as slugs, which are closed to mass transfer with each other and therefore act as batch reactors moving through space.

a shallow pond or lake that is well stirred by wind action can be modeled as a batch reactor. A pool in a low gradient stream, which is stirred by the stream flow, can be modeled as a mixed flow reactor; and a parcel of fluid flowing in a rock fracture can be modeled as a plug flow reactor. In addition to using chemical reactors models to understand natural situations, they are the basis for designing experiments to measure rates.

Chemical reactors can be embellished to provide effective mixing and to establish well-defined physical relationships among reacting phases. Figure 4.2 illustrates some examples of methods devised to bring solids into contact with solutions under controlled hydrodynamic conditions. Mixing by an overhead stirrer can suspend small particles to produce a homogeneous slurry. Magnetic stirrers can also be used for mixing, but the motion of the magnet on the reactor bottom can grind the particles, which changes the surface area of the solid to the mass of the solution in an unpredictable way. Shaking, rocking, or rotating a reactor that has a free space (gas phase) can also produce effective mixing. This scheme is especially useful for batch reactors that must be sealed to avoid the gain or loss of a vapor phase. Reactors containing a slurry, packed bed, or monolith can be mixed using an external recycle loop containing a pump that provides a mixing flow. If the loop is closed, the system is a batch reactor. If solution is added and removed from the loop, the system is a mixed flow reactor. Larger grains can be held in a packed bed through which the solution is forced by pumping. If the bed is long compared to its lateral dimensions, the reactor will perform as a plug

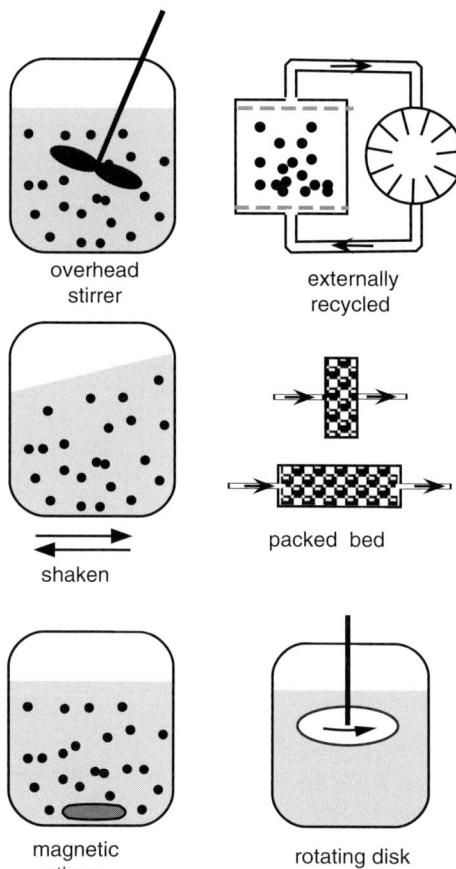

Figure 4.2. Examples of schemes bring solids into contact with fluids in chemical reactors.

flow reactor. If the bed is thin relative to its lateral dimensions, it will approximate the behavior of a mixed flow reactor. Monolith samples, which can be solid blocks of material or particles imbedded in an inert solid matrix, are desirable because they can be cleaved or ground and polished to present well-defined surfaces to the solution and these surfaces can be observed during and after their reaction with the solution. Monoliths can be presented to the solution in several ways ranging from simply suspending them in a stirred solution to flowing the solution past them using various schemes. A widely used scheme is to embed the monolith into the face of a disk, which is attached to a rotating shaft. This rotating disk exposes the solid to a well-defined hydrodynamic situation, which makes this apparatus useful for studying the rate of transport-limited dissolution (Alkattan et al., 1998; Levich, 1962; Sjöberg and Rickard, 1983). Many other schemes are possible, but regardless of the scheme, the reactor should be designed to operate as closely as possible to an ideal chemical reactor to simplify modeling its behavior.

Tracer kinetics

Tracer kinetics interpreted using the continuity equation is an effective way to characterize the mixing and flow behavior of chemical reactors. Tracers are inert substances that are added to reactors, and the tracer concentration in the reactor or in the effluent stream is used to infer something about the reactor performance. Ideal tracers are inert so they do not react with anything in the reactor. This means that the r_{gen} term in the continuity equation (4.1) equals zero. Tracers are frequently used to determine reactor dimensions or flow rates in field settings where direct physical measurements are impractical or impossible. In addition to displaying conservative behavior, field tracers must be environmentally benign, inexpensive, and easy to quantify at low concentrations. Bromide and lithium ions meet these criteria and are frequently used. The environmental and cost requirements for tracers used in laboratory, pilot scale, and industrial settings are less stringent because smaller amounts are used and the resulting solutions can be captured and treated before disposal. Dyes or even radioactive tracers might be used under these more controlled conditions. Natural situations are often complicated so the models in this section should be viewed as approximations of real systems. The models described by Rubinow (1975) illustrate how tracer models are applied to more challenging cases.

Tracer dosing follows one of two end member scenarios; the tracer is either added as an instantaneous spike or it is added continuously beginning at a known time. Adding spikes is the simplest method but if the reactor is large a single spike may not produce a detectable signal. On the other hand, continuous addition requires a metering pump and a larger amount of tracer. Serendipitous natural tracers can be used, but they are often introduced in unpredictable and irregular amounts. This makes them useful for tracing flow paths even if they cannot be used to develop quantitative models of reactor volumes or flow rates.

Tracer behavior is different in each of the ideal chemical reactors, so the first step in developing a tracer kinetics model is to decide which reactor type best simulates the real situation and which kind of tracer dosing has occurred. If the amount and timing of tracer introduction is known, a forward model can be developed. If the amount and timing of tracer detected is known, an inverse model can be produced. The equations derived in the following section use concentration units of mg/L because these units are typical of field studies. For laboratory experiments the models would use molal concentration units. The mass (M) variables would be replaced by mole quantities (n). The volume variables (V) would be replaced by mass of water (M) and the flow rate (Q) would have units of kg/sec.

Tracer spike addition to a batch reactor

Adding a tracer spike to a batch reactor produces a step increase in the tracer concentration in the reactor and the concentration remains constant thereafter. No tracer is discharged from the reactor and no tracer is generated in the reactor.

$$M_{acc} = M_{in} - M_{out} + M_{gen} = M_{in} - 0 + 0 = M_{in} \qquad (4.2)$$

The concentration in the reactor after the spike is added (C, mg/L) is the sum of the initial mass of tracer (M_o, mg) and the mass of tracer added in the spike (M_s, mg) divided by the total volume of solution (V_s, L).

$$C = \frac{M_o + M_s}{V_o + V_s} = \frac{C_o V_o + C_s V_s}{V_o + V_s} \qquad (4.3)$$

The initial amount of the tracer in the reactor is the original concentration of tracer (C_o, mg/L) times the original volume of solution (V_o, L).

$$M_o = C_o V_o \qquad (4.4)$$

The mass of the tracer in the spike is its concentration (C_s, mg/L) times the volume of the spike solution.

$$M_s = C_s V_s \qquad (4.5)$$

Example 4.1. Volume of a hot spring pool

Hot spring pools are usually well stirred by thermal convection and rising gas bubbles, so they can be modeled as ideal batch reactors and the volume of water in the pool can be determined using a tracer spike.

A typical spike might consist of 1 L of a 4.6×10^5 mg/L bromide solution. Once the spike is added to the hot spring, it takes a short time for it to mix throughout the spring pool. If the completely mixed pool water contains 4.6 mg/L, Eq. (4.3) can be rearranged to calculate the volume of the pool (V_o, L). This relationship can be further simplified by realizing that $C_s \gg C$ so C can be neglected in the numerator and that $C \gg C_o$ so that C_o can be neglected in the denominator.

$$V = V_s \left(\frac{C_s - C}{C - C_o} \right) \approx V_s \left(\frac{C_s}{C} \right) \qquad (4.6)$$

$$V \approx V_s \left(\frac{C_s}{C} \right) = 1\,\text{L} \left(\frac{4.6 \times 10^5 \frac{\text{mg}}{\text{L}}}{4.6 \frac{\text{mg}}{\text{L}}} \right) = 1 \times 10^5\,\text{L} \qquad (4.7)$$

The volume of the spring pool is 1×10^5 L.

Tracer spike addition to a mixed flow reactor

Adding a tracer spike to an ideal mixed flow reactor produces a step increase in the tracer concentration just as in the batch reactor case, but in this case the tracer's concentration declines over time as the reactor's effluent carries it away. If the amount of solution in the reactor remains constant and no more tracer is added by the feed solution or by generation in the reactor, the continuity equation simplifies to make the rate of accumulation equal to the rate of loss by the effluent flow.

$$r_{acc} = r_{in} - r_{out} + r_{gen} = 0 - r_{out} + 0 = -r_{out} \qquad (4.8)$$

The rate of removal of the tracer by the effluent solution is directly proportional to its flow rate (Q, L/sec) divided by the reactor volume (V, L). This quotient (k_Q, sec^{-1}) behaves like the rate constant in a first-order rate equation.

$$\frac{dC}{dt} = -\left(\frac{Q}{V}\right)C = -k_Q C \qquad (4.9)$$

Equation (4.9) can be rearranged and integrated from the initial tracer concentration (C_o, mg/L) to the tracer concentration at some later time (C, mg/L).

$$\int_{C_o}^{C} \frac{dC}{C} = -k_Q \int_0^t dt \qquad (4.10)$$

$$\ln\left(\frac{C}{C_o}\right) = -k_Q t \qquad (4.11)$$

This equation can be transformed to give the concentration of tracer in the reactor as a function of time.

$$C = C_o e^{-k_Q t} = C_o e^{-\left(\frac{Q}{V}\right)t} \qquad (4.12)$$

Example 4.2. Hot spring flow rate

Once the tracer concentration has homogenized in the hot spring, as described in Example 4.1, discharge from the spring carries away some of the tracer so the concentration in the spring pool declines as shown in Figure 4.3.

Equation (4.11) can be used along with the data from this graph to find the discharge of the hot spring. After 5 hours (18,000 sec) have passed, the concentration is 0.76 mg/L so $C/C_o = 0.165$ and $\ln(C/C_o) = -1.80$. This means that $k_Q = -1.8/-1.8 \times 10^4 = 1.0 \times 10^{-4}$ sec^{-1}. The volume of water in the spring pool is 1.0×10^5 L so the discharge $Q = k_Q V = (1.0 \times 10^{-4}$ sec$^{-1})(1.0 \times 10^5$ L$) = 10$ L/sec.

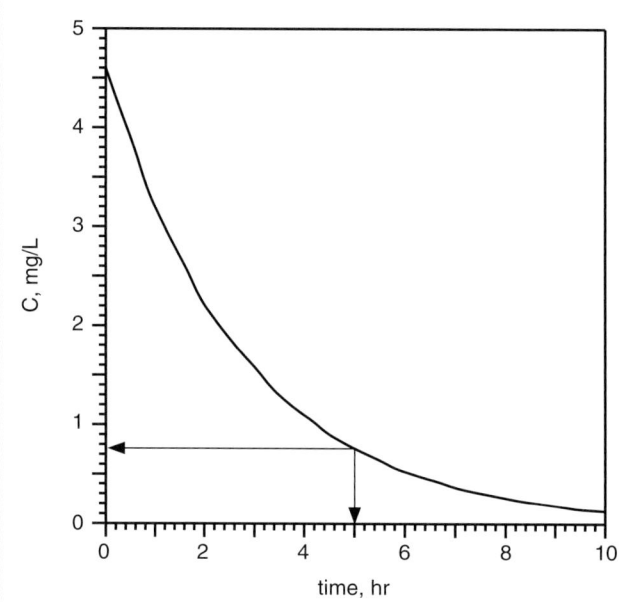

Figure 4.3. Concentration of tracer in a hot spring as a function of time. After 5 hours have passed, the tracer concentration in the spring pool is 0.76 mg/L.

Tracer spike addition to a plug flow reactor

Adding a tracer spike to an ideal plug flow reactor creates a slug from which no tracer is lost or gained as it traverses the length of the reactor. For an ideal plug flow reactor, the concentration of the tracer in the slug is the same regardless of its position in the reactor. In real plug flow reactors, diffusion and dispersion cause some of the tracer to advance ahead of the slug and some to lag behind the slug. The peak of tracer concentration is usually chosen as the location of the slug. The time needed for the slug to travel a given distance is used to characterize the flow path. For example, the time needed for a slug to pass through a packed bed can be used to find the volume of fluid in the packed bed if the flow rate (Q, L/sec) and bed porosity (ϕ) are known.

$$V = \frac{Qt}{\phi} \qquad (4.13)$$

If the reactor does not contain a packed bed, $\phi = 1$.

Plug flow reactor models must be constructed with a frame of reference in mind. An observer riding along on the tracer slug would see no change in concentration over the modeled time and could infer nothing about the reactor.

Continuous tracer addition to a batch reactor

If a solution containing a dissolved tracer is added to an ideal batch reactor at a constant rate, the rate of accumulation of both the tracer and the solution is described by the continuity equation.

$$r_{acc} = r_{in} - r_{out} + r_{gen} = r_{in} - 0 + 0 = r_{in} \qquad (4.14)$$

The mass of tracer in the reactor (M, mg) is the initial mass (M_o, mg) plus the mass added over the interval of tracer injection (ΔM, mg) and that amount equals the product of concentration in the tracer feed (C_{in}, mg/L) and the feed flow rate (Q_{in}, L/sec) times the time interval (t, sec).

$$M = M_o + \Delta M = M_o + C_{in} Q_{in} t \qquad (4.15)$$

The volume of solution in the reactor at the end of the tracer injection (V, L) is the initial volume (V_o, L) plus the volume added by the injection flow.

$$V = V_o + Q_{in} t \qquad (4.16)$$

The concentration of tracer (C, mg/L) in the reactor at time = t (sec) is the mass of tracer from Eq. (4.15) divided by the volume of solution from Eq. (4.16).

$$C = \frac{M}{V} = \frac{M_o + C_{in} Q t}{V_o + Q t} \qquad (4.17)$$

Continuous tracer addition to a mixed flow reactor

If a tracer is added to an ideal mixed flow reactor at a constant rate, the rate of accumulation of the tracer in the reactor amounts to the difference between the rate of input by the feed stream and the rate of removal in the effluent stream.

$$r_{acc} = r_{in} - r_{out} + r_{gen} = r_{in} - r_{out} + 0 = r_{in} - r_{out} \qquad (4.18)$$

$$\frac{dC}{dt} = \left(\frac{Q}{V}\right) C_{in} - \left(\frac{Q}{V}\right) C = k_Q (C_{in} - C) \qquad (4.19)$$

The concentration in the feed stream is constant so this equation can be rearranged and integrated.

$$\int_{C_o}^{C} \frac{dC}{C_{in} - C} = k_Q \int_{0}^{t} dt \qquad (4.20)$$

$$\ln\left(\frac{C_{in} - C_o}{C_{in} - C}\right) = k_q t \tag{4.21}$$

This equation can be rearranged to find the concentration of the tracer in the reactor as a function of time.

$$C = C_{in} - (C_{in} - C_o)e^{-k_q t} \tag{4.22}$$

Continuous tracer addition to a plug flow reactor

This case is very similar to the previously discussed case for the injection of a tracer spike. The time needed for the first spiked slug to reach the reactor's discharge is found by rearranging Eq. (4.13).

$$t = \frac{V\phi}{Q} \tag{4.23}$$

All subsequent slugs contain the same tracer concentration as the first.

Measuring reaction rates using chemical reactors

Rate measurements are best performed with experiments that closely approximate the behavior of ideal chemical reactors. The concept of ideal chemical reactors is well developed (Aris, 1989; Levenspiel, 1972a, 1972b; Rimstidt and Newcomb, 1993), so that the methods of extracting rates from the experimental data are relatively simple and reliable. Chemical reactors can have many different design elements, some of which are illustrated in Figure 4.2, but as long as their behavior approximates an ideal chemical reactor the following analysis methods are applicable.

Batch reactor experiments

Batch reactor experiments are easy to design and execute but are difficult to interpret. Because batch reactors are closed systems, the rate of accumulation of a species in a batch reactor equals that species' rate of generation. The concentration versus time data from batch reactor experiments must be manipulated to extract the rate of generation. If an elementary reaction is responsible for the rate of generation, the rate equation will contain a concentration term raised to an integer power and it is possible to identify an appropriate integrated rate equation by fitting the concentration versus time data to selected integrated rate equations until a good match is found. This guess method is often impractical because there are innumerable choices for the integrated rate equation. Furthermore, fits of concentration versus time data to integrated rate equations, either using linearized forms or using non-linear regression, will produce dubious results unless the data span a very large extent of reaction (Rimstidt and Newcomb, 1993).

A more common approach to analyzing batch reactor data is to determine rates by taking the derivative of the concentration versus time data. If the concentration is determined at equally spaced times, the numerical derivative of the concentration versus time data can be used to find the rate of generation at each sampling time. A rate equation can then be developed by correlating the rate at each sample time with the conditions that existed at that time. Single experiments seldom span a large enough extent of reaction to produce a reliable correlation, so rates from multiple experiments that span a large range of concentrations must be pooled.

An alternative strategy is to perform numerous short-term experiments that span a wide range of initial conditions and then use the derivative of a polynomial fit of the concentration versus time data for each experiment to find the rate at the beginning of each experiment. An advantage of finding this initial rate is that all the rate-controlling parameters are fully constrained by the experimental design at the beginning of the experiment. This method finds the derivative, at $t = 0$, of an arbitrary function that fits the concentration versus time data. The simplest fitting function is a straight line and the slope of the line equals the rate. This works for cases where the rate is nearly constant over the duration of the experiment. If the rate changes significantly during the experiment, the concentration versus time graph will be curved and a straight line underestimates the rate. Fitting the data to a second (or higher) order polynomial will accommodate some curvature.

$$m = a + bt + ct^2 \qquad (4.24)$$

The rate is approximated by the derivative of Eq. (4.24).

$$\frac{dm}{dt} = b + 2ct \qquad (4.25)$$

The initial rate ($t = 0$) is b. When $t = 0$, the rate-controlling parameters are known so this rate can be correlated with those conditions. This method is quite effective if the extent of the reaction is small ($<\sim$5% of the reactant consumed) otherwise Eq. (4.25) will systematically underestimate the initial rate. A higher-order polynomial will better fit data for greater extents of reaction, but that approach is not recommended because it sometimes causes unusual flexing of the fit near $t = 0$ leading to an incorrect estimate of the initial rate.

An alternative to the polynomial method is the chord method. The chord method is recommended for experiments that reach larger extents of reaction (Casado *et al.*, 1986; Piscitelle, 1990; Waley, 1981). This method extracts the initial rate from the slopes, $(m_t - m_o)/\Delta t$, of chords drawn between each concentration (m_t) point and the starting concentration (m_o). The slopes of these chords are graphed versus time and this function is extrapolated to $t = 0$. This gives the slope of the tangent to the concentration versus time graph at $t = 0$, which is the initial rate. Graphs of the chords' slopes versus time are nearly linear for small extents of reaction but do show some curvature, which can be fit by a second-order polynomial. The initial rate is found by solving this polynomial for the slope at $t = 0$.

Example 4.3. Initial rate of scorodite dissolution using the polynomial and chord method

Harvey et al. (2006) measured the rate of scorodite ($FeAsO_4 \cdot 2H_2O$) dissolution using a batch reactor. One of the experiments produced the concentration of arsenic versus time data given in Table 4.1. These data can be analyzed to find the initial rate using either the polynomial fit method or the chord method.

Table 4.1. *Concentration versus time data for a scorodite dissolution rate experiment*

t	m	$\Delta m/\Delta t$
sec	molal	molal/sec
0	1.89×10^{-6}	
900	2.95×10^{-6}	1.19×10^{-9}
1800	3.81×10^{-6}	1.07×10^{-9}
3600	4.95×10^{-6}	8.50×10^{-9}
7200	5.77×10^{-6}	5.39×10^{-9}
10800	6.34×10^{-6}	4.12×10^{-9}
14400	7.12×10^{-6}	3.63×10^{-9}

The heavy line in Figure 4.4 is a second-order polynomial fit of all the concentration versus time data in Table 4.1.

$$m_{As} = -2.59 \times 10^{-14} t^2 + 6.89 \times 10^{-10} t + 2.34 \times 10^{-6} \tag{4.26}$$

The "b" term for this fit is 6.89×10^{-10}. A comparison of this fit to the first few data points shows that the slope of the line at $t = 0$ is too small. This problem can be remedied by fitting only the first three data points to produce the light line shown in Figure 4.4.

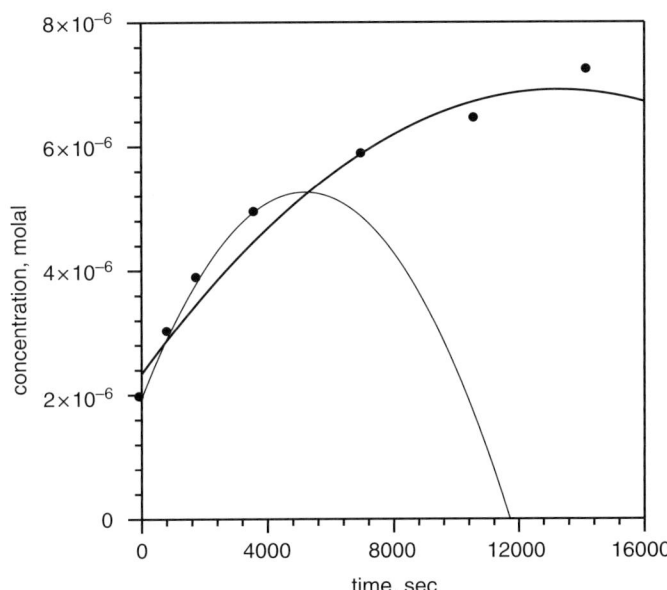

Figure 4.4. Concentration of arsenic released by scorodite dissolution versus time fit to a second-order polynomial. The heavy line is a fit to all the data and the light line is a fit to the first three data.

$$m_{As} = -1.32 \times 10^{-13} t^2 + 1.31 \times 10^{-9} t + 1.89 \times 10^{-6} \qquad (4.27)$$

The "b" term for this fit is 1.31×10^{-9}, which is nearly twice the value obtained by fitting all the data points. This example illustrates that a relatively large amount of error can result from fitting data for too large an extent of reaction.

The chord method is an improvement over the polynomial fit method but it requires slightly more manipulation of the data. First the slopes of the chords shown in Figure 4.5a are calculated and tabulated. Then the chord slopes are graphed versus time as shown in in Figure 4.5b and these points are fit to a second-order polynomial.

$$s = 5.56 \times 10^{-18} t^2 + 1.45 \times 10^{-13} t + 1.31 \times 10^{-9} \qquad (4.28)$$

When $t = 0$, the slope of the chord, which approximates the tangent, is 1.31×10^{-9} molal/sec, which is the initial rate.

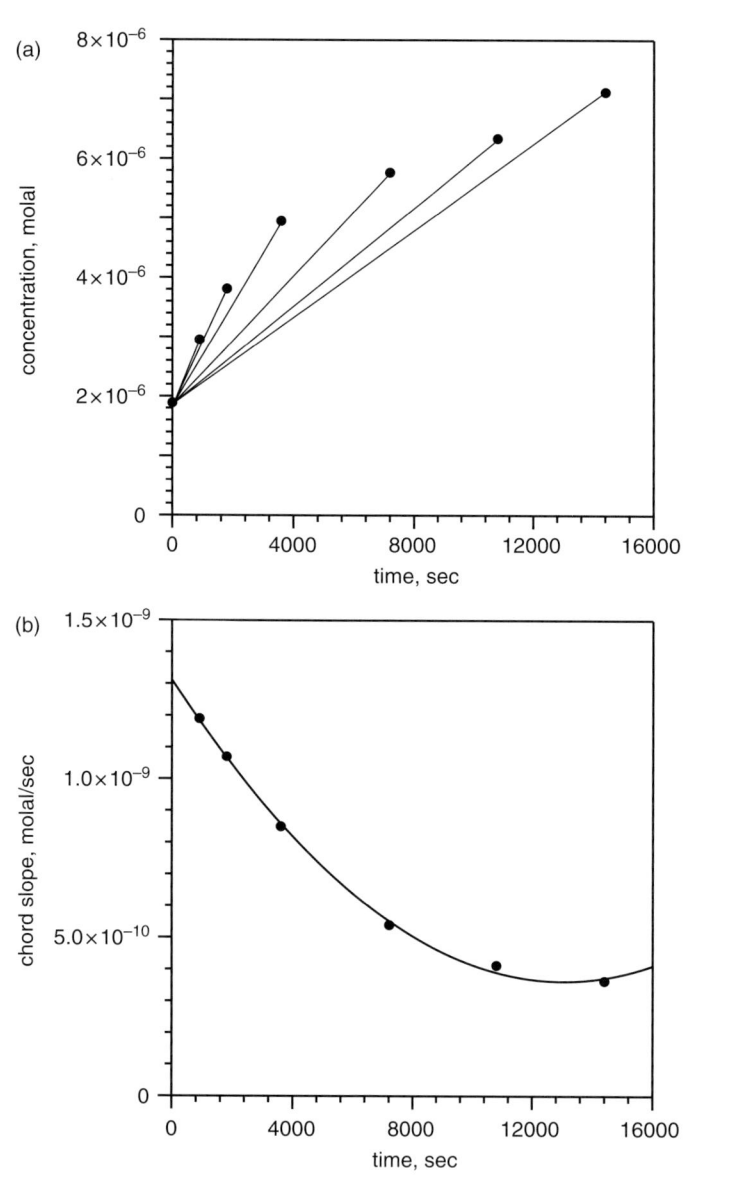

Figure 4.5. (a) Concentration of arsenic versus time graph with chords connecting the concentration at zero time and the concentration at the time of each sampling. (b) A graph of the slope of the chords versus time. Extrapolation of this graph to zero time gives slope of the chord at $t = 0$, which is the initial rate.

Mixed flow reactor experiments

Mixed flow reactor experiments are relatively difficult to design and execute but they produce data that are easy to analyze. When the experiment reaches steady state, the concentration of all species in the reactor is constant. This means that the rate of species accumulation is zero, so the rate of generation is the difference between the input rate, which is the flow rate (Q, kg/sec) multiplied by the feed concentration (m_{in}, mol/kg), and the output rate, which is the flow rate multiplied by the effluent concentration (m_{out}, mol/kg).

$$r_{gen} = r_{out} - r_{in} = (m_{in} - m_{out})Q \tag{4.29}$$

Usually m_{in} is set by the experimental design and remains constant so only the concentration in the effluent needs to be determined. It is advisable to determine the flow rate (Q, kg/sec) at each sample time by weighing the sample container before and after a sample is taken for chemical analysis because even the best pumps do not maintain constant flow rates over long time intervals. Weighing samples is more precise and more accurate than volumetric measurements. If the reactor is well stirred, the effluent stream is a representative sample of the solution in the reactor so its analysis establishes the chemical conditions that produced the observed rate.

> **Example 4.4.** Rate consumption of Fe^{3+} by reaction with pyrite in a MFR
>
> Rimstidt and Newcomb (1993) performed an experiment to measure the rate of pyrite reaction with ferric iron in a mixed flow reactor. The reactor, containing 2 g of pyrite with a specific area of 0.047 m²/g, was fed by a 1.0×10^{-3} m Fe^{3+} solution at a flow rate of 2.70×10^{-5} kg/sec. After a time, the experiment reached steady state where the concentration of Fe^{3+} in the effluent solution maintained a constant value of 1.86×10^{-4} molal.
>
> Based on this information, the rate of Fe^{3+} consumption in the reactor was found using Eq. (4.29).
>
> $$r_{gen} = (m_{in} - m_{out})Q = (10.0 \times 10^{-4} - 1.86 \times 10^{-4})(2.70 \times 10^{-5}) = 2.20 \times 10^{-8} \frac{mol}{sec}$$
>
> The flux of Fe^{3+} to the surface of the pyrite was determined by dividing this rate by the total surface area of the pyrite.
>
> $$J_{Fe^{3+}} = \frac{r_{Fe^{3+}}}{A} = \frac{2.20 \times 10^{-8} \frac{mol}{sec}}{(2\ g)\left(0.047 \frac{m^2}{g}\right)} = 2.34 \times 10^{-7} \frac{mol}{m^2 sec} \tag{4.30}$$

Plug flow reactor experiments

Plug flow reactor experiments are more complicated to design and execute than batch reactor experiments, and they produce concentration versus time data that must be analyzed using the same methods that are used for batch reactors. As a result, plug flow reactor experiments are seldom used to determine rates. However, the very high surface area to solution volume of packed beds achievable in plug flow reactors makes them useful for measuring the dissolution or precipitation rates of very unreactive solids. A key assumption for such experiments is that the surface reactivity of the solid does not change during the experiment.

Either the polynomial method or the chord method can be used to find an initial rate from plug flow reactor data. The chord method is especially convenient in this case because the difference between the concentration in the effluent and feed streams (Δm, molal) divided by the transit time for a slug of solution is a chord. The transit time is the reactor volume (V, L) times the porosity (ϕ, no units) divided by the flow rate (Q, L/sec).

$$\Delta t = \frac{V\phi}{Q} \tag{4.31}$$

Setting Q to different values produces different transit times with corresponding different values of Δm. Dividing the resulting Δm by the transit time gives the slope of a chord. Extrapolating a graph of these chord slopes to $\Delta t = 0$ gives the initial rate.

The chord method determines the initial rate, R (molal/sec). To convert this rate into the flux of species to or from the surface of the solid, R must be divided by the A/M ratio for the packed bed as described in Chapter 3.

Example 4.5. Rate consumption of Fe^{3+} by reaction with pyrite in a PFR

Rimstidt and Newcomb (1993) performed an experiment in a plug flow reactor where a feed solution containing 1.0×10^{-3} m Fe^{3+} passed through a packed bed of pyrite with an A/M of 334 m²/kg at different flow rates. The data for this experiment are shown in Table 4.2.

Figure 4.6 shows that regression of $\Delta m/\Delta t$ versus Δt to a second-order polynomial gives an excellent fit with $R^2 = 0.997$.

The chord slopes given in the table were correlated with transit time using a second-order polynomial (Figure 4.6).

$$\frac{\Delta m}{\Delta t} = 6.25 \times 10^{-8} t^2 - 2.23 \times 10^{-6} t + 2.42 \times 10^{-5} \tag{4.32}$$

Table 4.2. *Data for Fe^{3+} reaction with pyrite plug flow experiment.*

Q	m_{out}	Δt	Δm	$\Delta m/\Delta t$
L/sec	molal	sec	molal	molal/sec
3.92×10^{-5}	3.94×10^{-6}	20.50	9.61×10^{-5}	4.69×10^{-6}
8.83×10^{-5}	1.56×10^{-5}	9.10	8.44×10^{-5}	9.27×10^{-6}
1.28×10^{-4}	2.19×10^{-5}	6.30	7.81×10^{-5}	1.24×10^{-5}
1.71×10^{-4}	3.06×10^{-5}	4.70	6.94×10^{-5}	1.48×10^{-5}
2.17×10^{-4}	3.66×10^{-5}	3.70	6.34×10^{-5}	1.71×10^{-5}

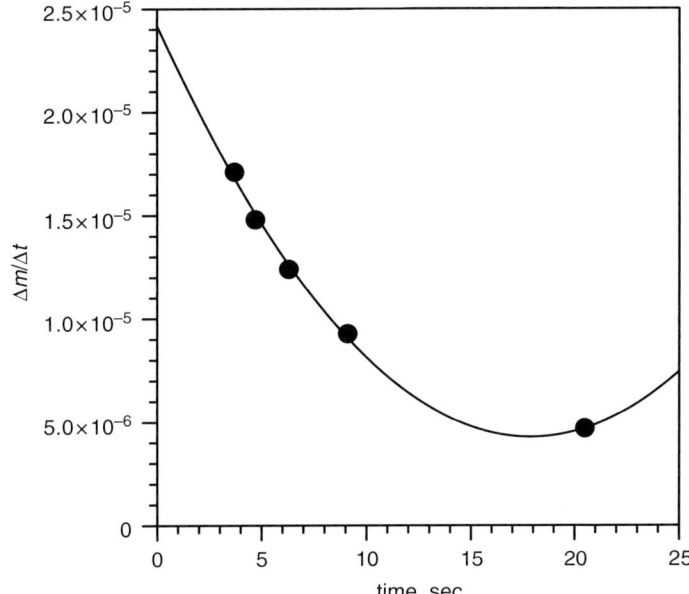

Figure 4.6. Slope of the chords versus time for the Fe^{3+} reaction with pyrite in a plug flow reactor experiment. The intercept of this graph is the initial rate.

Extrapolation of this function to $\Delta t = 0$ gives an initial reaction rate of 2.42×10^{-5} molal/sec. To obtain the flux of Fe^{3+} from the solution to the pyrite surface this rate is divided by the A/M for the experiment to get $J = 6.72 \times 10^{-8}$ mol/m²sec.

Double log method

To produce geochemical rate models, rates determined by reactor experiments must be converted into rate equations that summarize how the rate varies with solution composition, temperature, and other rate-determining variables. If the rates are determined at near-equilibrium conditions, the rate data must be fit to an equation that takes into account both the forward and reverse rate. Most geochemical rate experiments are designed to measure rates for far-from-equilibrium conditions where the reverse reaction rate is effectively zero. These experimental rates can be fit to a simple equation that relates the rate to the product of the concentration (m, molal) of each reacting species raised to a power (n).

$$r = k \prod m_i^{n_i} \tag{4.33}$$

This equation is nonlinear; direct fitting requires the use of nonlinear regression, which can be a daunting task. This problem can be avoided by log-transforming Eq. (4.33) to a linear form.

$$\log r = \log k + \sum n_i \log m_i \tag{4.34}$$

Fitting the data to this equation using linear regression is a simple way to find the rate constant and the apparent reaction orders. This equation can be further generalized by transforming the $\log k$ term using the Arrhenius equation (see Chapter 5). The activation energy (E_a, kJ/mol) term in this equation expresses how the rate changes with changing temperature (T, K).

$$k = Ae^{-\left(\frac{E_a}{R}\right)\frac{1}{T}} \tag{4.35}$$

A is a pre-exponential factor with the same units as the rate constant and R is the gas constant (8.314 J/mol K). Equation (4.35) can be log-transformed to express $\log k$ in terms of $1/T$.

$$\log k = \log A - \left(\frac{E_a}{2.303R}\right)\frac{1}{T} \tag{4.36}$$

Combining Eqs (4.34) and (4.36) produces a function that relates the observed rate to temperature and solution composition.

$$\log r = \log A - \left(\frac{E_a}{2.303R}\right)\frac{1}{T} + \sum n \log m_i \tag{4.37}$$

Multiple linear regressions must be used to fit data to Eq. (4.37). See for example Rimstidt *et al.* (2012).

Non-ideal chemical reactors

Example 4.6. Fluorapatite dissolution flux as a function of pH

Chaïrat et al. (2007) measured the rate of calcium release from dissolving fluorapatite over the pH range of 3 to 7. The predominant reaction produces calcium and fluoride ions and undissociated phosphoric acid.

$$Ca_5(PO_4)_3F + 9\,H^+ = 5\,Ca^{2+} + 3\,H_3PO_4 + F^- \tag{4.38}$$

Figure 4.7 shows that the logarithm of the Ca^{2+} release flux versus pH is linear over this pH range so these data can be fit to Eq. (4.34).

$$\log J = -4.91(0.09) - 0.88(0.02)pH \tag{4.39}$$

This equation can be transformed to give the Ca^{2+} flux as a function of the activity of H^+.

$$J = 10^{-9.91(0.09)} a_{H^+}^{0.88(0.020)} \tag{4.40}$$

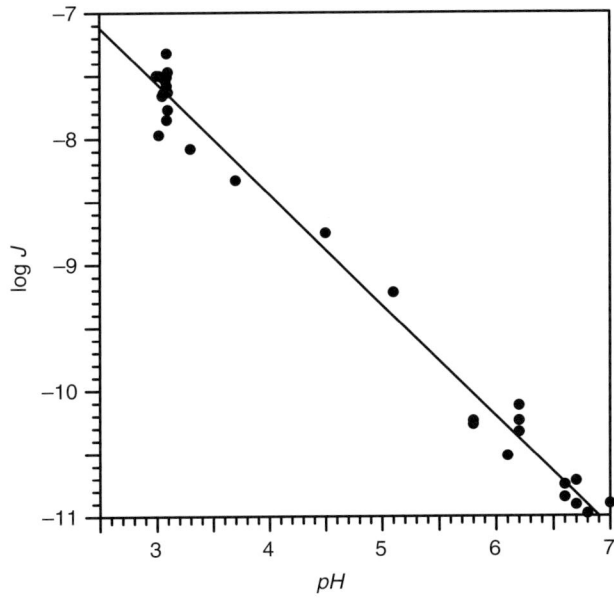

Figure 4.7. Logarithm of the Ca^{2+} release flux versus pH for fluorapatite dissolution. The equation for the best-fit line is (4.39).

Non-ideal chemical reactors

There are many cases when chemical reactors behave non-ideally so that reactor models must be adjusted to account for the processes that cause the non-ideal behavior. Following are a few examples of how non-ideal cases can be modeled.

Sampled batch reactor

Most batch reactor experiments that are used to measure the dissolution of solids are designed so that the mass of solution declines as samples are removed but the amount of solid, which settles to the bottom of the reactor during the sampling, remains constant. This means that the A/M ratio changes from sample to sample. To correct the concentration of each sample to the concentration that would have been found if the A/M ratio had remained constant, the concentration of each subsequent sample must be adjusted downward to account for the declining mass of solution. The amount of a substance dissolved into the solution during an interval between samples equals the change in the concentration over that interval multiplied by the mass of solution in the reactor during that interval. The total amount of substance (n, moles) that has been released to the solution at any sample time is the sum of the mass of solution (M, kg) times the change in concentration between samples (Δm, mol/kg).

$$n = \sum_{\text{sample 1}}^{\text{sample } n} (\Delta m) M \qquad (4.41)$$

Dividing n at each sample time by the initial mass of solution in the reactor corrects the concentration to what it would have been if no solution had been withdrawn by previous samples. This corrects the concentration versus time data to match that expected for an ideal batch reactor.

Alternatively, using the chord or polynomial fit method to find the initial rate approximately corrects for the changing A/M ratio in a sampled batch reactor provided the sample size is small relative to the total amount of solution in the reactor.

Mixing in reactors

Modeling incomplete mixing in chemical reactors can be quite challenging. Most models use a distribution function of the residence time for fluid elements in the reactor (Danckwerts, 1953; Denbigh and Turner, 1984). Nauman (2008) provides a concise history and a useful review of this theory.

Many reaction-transport models treat parcels of fluid as if they are slugs flowing in a plug flow reactor. In real systems, hydrodynamic mixing and diffusion (Aris, 1956; Taylor, 1953) cause dispersion of species among the slugs. Gelhar et al. (1992) review the factors that affect dispersion in aquifer systems.

Non-steady state mixed flow reactors

Reaction rates can be determined from mixed flow reactors operating under non-steady state conditions using the following procedure. Material balance as defined by Eq. (4.1) requires that the rate of generation of a species by a reaction (r_{gen}, mol/sec) equals the rate of output (r_{out}, mol/sec) minus the rate of input (r_{in}, mol/sec) plus the rate of accumulation (r_{acc}, mol/sec) in the reactor.

$$r_{gen} = (r_{out} - r_{in}) + r_{acc} \qquad (4.42)$$

The ($r_{out} - r_{in}$) term is the species concentration in the feed solution (m_{in}, mol/kg) minus its concentration in the effluent (m_{out}, mol/kg) multiplied by the flow rate of the effluent stream (Q, kg/sec).

$$r_{out} - r_{in} = (m_{out} - m_{in})Q \qquad (4.43)$$

The rate of accumulation of the species in the reactor is the time rate of change of its concentration multiplied by the mass of solution in the reactor (M, kg).

$$r_{acc} = \left(\frac{dm_{out}}{dt}\right)M \qquad (4.44)$$

Equations (4.42), (4.43), and (4.44) are combined to find the rate of reaction for non-steady state conditions.

$$r_{gen} = (m_{out} - m_{in})Q + \left(\frac{dm_{out}}{dt}\right)M \qquad (4.45)$$

The first term in this equation is simple to evaluate using the concentration and flow rate data that are typically collected for MFR experiments. The second term is evaluated by fitting the effluent concentration to a power law function and then taking its derivative.

$$m_{out} = at^b \qquad (4.46)$$

$$\left(\frac{dm_{out}}{dt}\right) = abt^{b-1} \qquad (4.47)$$

The a and b parameters are then used to find the rate of change in the effluent concentration. The rate of generation at each sample time is found by combining Eqs (4.45) and (4.47).

$$r_{gen} = (m_{out} - m_{in})Q + (abt^{b-1})M \qquad (4.48)$$

Equation (4.44) shows that the larger the mass of solution in the reactor the larger the rate of accumulation relative to the rate of generation and the longer it takes for the reactor to reach steady state. This means that a reactor designed to contain a very small amount of solution will reach steady state relatively quickly.

Example 4.7. Ca^{2+} flux from dissolving wollastonite in a non-steady state MFR

Weissbart and Rimstidt (2000) measured the concentration of Ca^{2+} in the effluent of a mixed flow reactor in which wollastonite was dissolving. In their experiments, the Ca^{2+} concentration in the effluent decreased continuously over the

Table 4.3. *Calculation of non-steady state rates of Ca^{2+} release from dissolving wollastonite. The mass of solution in the reactor 0.047 kg. Data from Weissbart and Rimstidt (2000).*

	Q	t	m_{Ca}	$(Q)(m_{Ca})$	$(dm_{Ca}/dt)(M)$	r_{Ca}
	kg/sec	sec	mol/kg	mol/sec	mol/sec	mol/sec
U1	4.25×10^{-5}	1632	1.43×10^{-3}	6.08×10^{-8}	-8.65×10^{-8}	-2.57×10^{-8}
U2	2.47×10^{-5}	15960	8.13×10^{-4}	2.01×10^{-8}	-1.02×10^{-9}	1.91×10^{-8}
U3	1.70×10^{-5}	83220	2.16×10^{-4}	3.67×10^{-9}	-4.08×10^{-11}	3.63×10^{-9}
U4	1.17×10^{-5}	97200	1.99×10^{-4}	2.32×10^{-9}	-3.02×10^{-11}	2.30×10^{-9}
U5	4.02×10^{-5}	99600	3.40×10^{-5}	1.37×10^{-9}	-2.88×10^{-11}	1.34×10^{-9}
U6	3.10×10^{-5}	159900	5.79×10^{-5}	1.79×10^{-9}	-1.14×10^{-11}	1.78×10^{-9}
U7	9.67×10^{-5}	161700	2.20×10^{-5}	2.13×10^{-9}	-1.12×10^{-11}	2.12×10^{-9}
U8	1.05×10^{-4}	177060	1.85×10^{-5}	1.95×10^{-9}	-9.38×10^{-12}	1.93×10^{-9}
U9	8.43×10^{-5}	346020	1.94×10^{-5}	1.64×10^{-9}	-2.54×10^{-12}	1.63×10^{-9}
U10	7.95×10^{-5}	509700	1.94×10^{-5}	1.54×10^{-9}	-1.20×10^{-12}	1.54×10^{-9}
U11	2.25×10^{-4}	600000	4.80×10^{-6}	1.08×10^{-9}	-8.70×10^{-13}	1.08×10^{-9}

course of the experiment and appears to not reach a steady state. The Ca^{2+} flux from the dissolving wollastonite can be found using Eq. (4.48) and the data in Table 4.3.

The $(Q)(m_{Ca})$ column of Table 4.3 is the rate that the effluent removes Ca from the reactor. It is the first term of Eq. (4.48). The second term in Eq. (4.48) is found by fitting the concentration of Ca^{2+} in the effluent versus time to a power law (Figure 4.8a).

$$m_{Ca^{2+}} = 3.52 t^{-0.948} \tag{4.49}$$

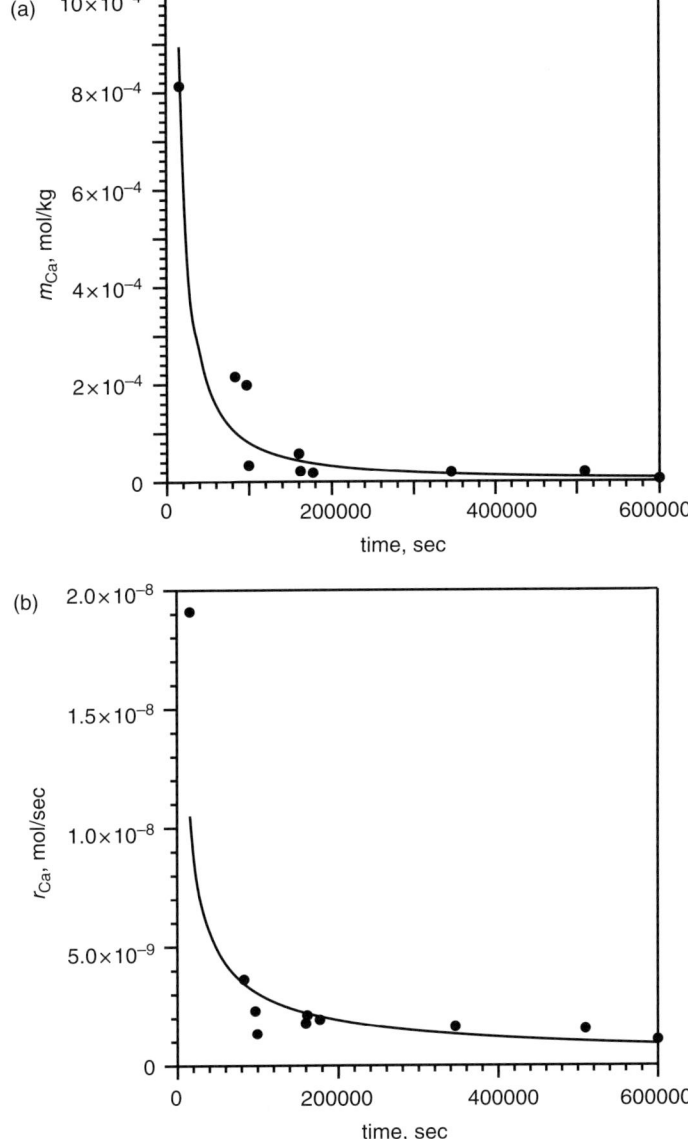

Figure 4.8. (a) Ca concentration in the MFR effluent solution as a function of time. (b) Rate of Ca release from dissolving wollastonite as a function of time. The equation for the line is $r_{Ca} = 7.18 \times 10^{-6} t^{0.675}$. A combination of errors caused the method to overcorrect the rate for sample U1 so that datum is not shown on the graphs.

The time rate of change of Ca concentration in the reactor solution is the derivative of Eq. (4.49).

$$\left(\frac{dm_{Ca}}{dt}\right) = (3.52)(-0.948)t^{-1.948} = -3.34t^{-1.948} \qquad (4.50)$$

Equation (4.50) is evaluated for each sample time and the $(dm_{Ca}/dt)(M)$ column reports the dissolved Ca that is retained in the reactor causing the Ca concentration to increase. Adding the $(Q)(m_{Ca})$ and $(dm_{Ca}/dt)(M)$ columns gives the rate of calcium release from the wollastonite at each sampling time (Figure 4.8b).

Minerals usually dissolve incongruently during the initial stages of the dissolution process. In this case, Ca is released from wollastonite faster than Si. In addition, the rate of both Ca and Si declines with increasing extent of reaction. Eventually the difference between release rates, as well as the change in the rates, from one sample time to the next becomes smaller than the resolution of the rate measurements. When both of these conditions exist the rates are assumed to represent the dissolution rate that is appropriate for long-term geochemical processes.

Chapter 5
Molecular kinetics

Rate equations are quantitative models of the time course of chemical reactions. Although rate equations are based on macroscopic observations, they reflect processes that occur at the molecular scale. This chapter reviews some of the important models that link these two scales. These models are especially useful because they constrain the mathematical form of rate equations and they provide a conceptual basis for thinking about the reactions. Because water is so important in geochemical systems, this chapter focuses on models for reaction rates in the aqueous phase.

Transition-state theory

Chemical reactions break and reform bonds within and between molecules so that the bonding arrangement of the reactants gives way to the bonding arrangement of the products. For a reaction to occur: (1) the molecules must "collide" with each other to form a cluster; (2) the atoms in that cluster must be configured in the approximate geometry of the products; and (3) the cluster must contain sufficient energy to allow the rearrangement of the electrons from the breaking bonds to the developing bonds.

The idea that molecules must collide to form a cluster correctly predicts that the reaction rate increases with concentration (or pressure) because increasing the number of molecules per unit volume will result in more collisions. This is the reason that rate equations stipulate that rates increase with increasing concentration, pressure, or activity.

The Franck–Condon principle states that the rearrangement of electrons to form new bonds occurs so quickly that the nuclei of the reacting atoms do not change position during the bond breaking–bond forming process. This means that for a reaction to occur, the reacting species must collide in such a way that they form a cluster with the atoms arranged with the geometry of the product molecules. This requirement means that most collisions do not lead to a reaction. A reactive cluster that meets these geometric constraints is called an activated complex.

Figure 5.1. The Maxwell–Boltzmann distribution of molecular energies shows that at low temperature there are fewer high-energy molecules available to form clusters that have energies that exceed the activation energy. Increasing the temperature increases the number of molecules that have enough energy to form activated complexes.

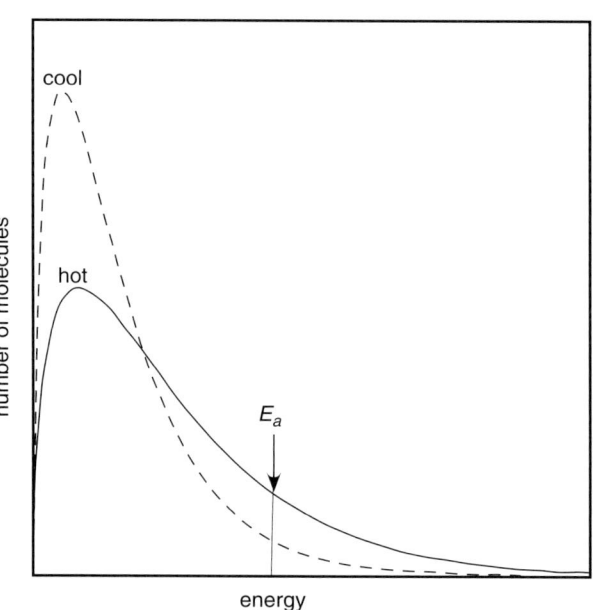

To form an activated complex, the colliding molecules must have enough energy to overcome the repulsive forces that would otherwise keep them from aligning into the geometry of the product molecules. The Maxwell–Boltzmann energy distribution portrays the sharing of energy among a collection of interacting molecules (Figure 5.1). This distribution shows that only a relatively few molecules have enough energy to form a cluster with the critical amount of energy needed for a reaction to occur. This critical amount of energy is called the activation energy (E_a, kJ/mol). Increasing the temperature distributes more thermal energy among the molecules so that there are more clusters with sufficient energy to form an activated complex. This explains why reaction rates increase with increasing temperature.

Transition-state theory, invented by Eyring (1935) and Evans and Polanyi (1935), is a statistical mechanics model that quantitatively accounts for the three factors that control the reaction process. Eyring used the term "activated complex" and Evans and Polanyi used the term "transition state", to describe the cluster of molecules that is transitional between the reactants and products. These terms continue to be used interchangeably.

Transition-state theory arose from statistical mechanics and describes the distribution of energy among molecules using partition functions. This means that a firm grasp of the theoretical underpinnings of transition-state theory requires an understanding of statistical mechanics. Houston (2001) describes many of the concepts discussed here from a statistical mechanics point of view and Fueno (1999) explains partition functions. However, the quasi-equilibrium model presented here is a simpler approach that illustrates the conceptual basis of transition-state theory without resorting to the use of partition functions.

Quasi-equilibrium model

The quasi-equilibrium model considers a simple reaction where atom A reacts with molecule BC and displaces atom C to produce molecule AB.

$$A + BC = AB + C \tag{5.1}$$

For this reaction to occur, molecule A must approach molecule BC from the correct direction with enough translational energy to form a transition-state cluster.

$$A + BC = ABC^* \tag{5.2}$$

Figure 5.2 shows how the energy of A + BC changes as a function of the distance between the molecules. A transition state can only form when the translational energy of A and BC equals or exceeds the energy needed to reach the saddle point that lies between the two valleys that parallel the A↔BC and the AB↔C axes. Once a transition-state cluster has formed it will break down to form AB + C.

The quasi-equilibrium model assumes that the transition-state cluster is in equilibrium with the A and BC species so that their activities are related by an equilibrium constant.

$$K^* = \frac{a_{ABC^*}}{a_A a_{BC}} \tag{5.3}$$

The activity of the transition-state cluster, ABC*, is found from the product of the equilibrium constant and the activity of each reactant.

$$a_{ABC^*} = K^* a_A a_{BC} \tag{5.4}$$

The concentration of ABC* is its activity divided by its activity coefficient.

$$m_{ABC^*} = \left(\frac{1}{\gamma_{ABC^*}}\right) K^* a_A a_{BC} \tag{5.5}$$

In this case, the concentration of the activated complex is expressed in mol/kg (molal) but other concentration units could be used. The rate that ABC* dissociates to form products, AB + C, is controlled by the vibrational frequency ($k_B T/h$) of the A–B bond.

$$R = \left(\frac{k_B T}{h}\right) m_{ABC^*} = \left(\frac{k_B T}{h}\right)\left(\frac{K^*}{\gamma_{ABC^*}}\right) a_A a_{BC} \tag{5.6}$$

82   Molecular kinetics

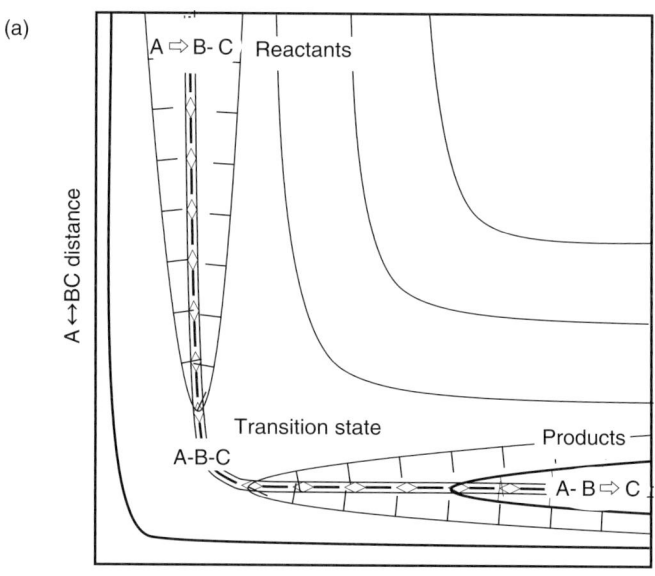

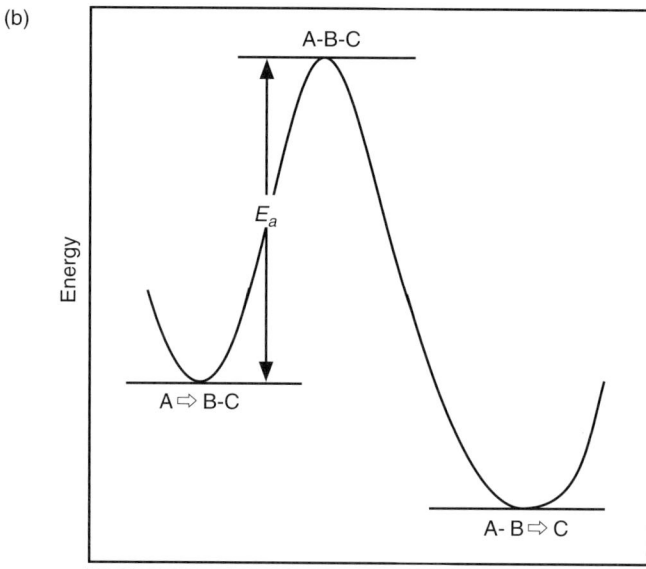

Figure 5.2. (a) Schematic contour plot showing the energy of the A-B-C cluster as a function of the distances between the atoms. ABC is the transition state. (b) Graph of the energy of the cluster as a function of its location along the reaction coordinate, which is the dashed line in (a).

The Boltzmann constant (k_B) has a value of 1.381×10^{-23} J/K and the Plank constant (h) is 6.626×10^{-34} J sec. This equation is further modified by introducing a transmission coefficient (κ, no units) to account for the fact that a few of the transition-state clusters fail to form products and instead revert back to being reactants. The transmission coefficient often has a value near one and is usually ignored.

$$R = \left(\frac{\kappa}{\gamma_{ABC^*}}\right)\left(\frac{k_B T}{h}\right) K^* a_A a_{BC} \tag{5.7}$$

Unit analysis of this equation shows that it has units of molal/sec. The first three terms are combined into a rate constant.

$$k = \left(\frac{\kappa}{\gamma_{ABC^*}}\right)\left(\frac{k_B T}{h}\right) K^* \tag{5.8}$$

The resulting equation is a rate law for a second-order reaction.

$$R = k a_A a_{BC} \tag{5.9}$$

Example 5.1. Concentration of the activated complex for the ferric thiosulfate reaction

Williamson and Rimstidt (1993) reported the rate of decomposition of ferric thiosulfate to form ferrous iron and tetrathionate ions. The reaction brings two ferric thiosulfate molecules together into an activated complex, which decomposes to form two ferrous ions and a tetrathionate ion.

$$2\ FeS_2O_3^+ = [(FeS_2O_3)_2^{2+}]^* \rightarrow 2\ Fe^{2+} + S_4O_6^{2-} \tag{5.10}$$

At 25°C the rate equation for this reaction is

$$R = -100 m^2_{FeS_2O_3^+} \tag{5.11}$$

If the concentration of $FeS_2O_3^+$ is 0.001 molal, the rate of disappearance of ferric thiosulfate is 0.0001 mol/kg sec. Equation (5.6) can be used to find the concentration of the postulated activated complex, $[(FeS_2O_3)_2^{2+}]^*$. At 298 K, $(k_B T/h) = 6.214 \times 10^{12}$ sec^{-1}.

$$m_{[(FeS_2O_3)_2^{2+}]^*} = \frac{R}{(k_B T/h)} = \frac{1 \times 10^{-4}}{6.21 \times 10^{12}} = 1.61 \times 10^{-19} \frac{mol}{kg} \tag{5.12}$$

This concentration is well below detection limits for aqueous species and any $[(FeS_2O_3)_2^{2+}]^*$ cluster that forms has a lifetime of ~1.6×10^{-13} sec, which makes this exercise entirely hypothetical but hopefully instructive.

Effect of temperature on rates

The quasi-equilibrium model can be further developed to show how temperature affects rates. The first step is to relate the free energy of activation (ΔG_r^*) to $\ln K^*$.

$$\Delta G_r^* = -RT \ln K^* \tag{5.13}$$

Because $\Delta G_r = \Delta H_r - T\Delta S_r$, Eq. (5.13) can be expanded to relate the enthalpy and entropy of activation to K^*.

$$\Delta H_r^* - T\Delta S_r^* = -RT \ln K^* \tag{5.14}$$

Combining Eqs (5.8) and (5.14) gives a relationship between the enthalpy and entropy of activation and the rate constant (k).

$$\ln k = \left(\frac{-\Delta H_r^*}{R}\right)\frac{1}{T} + \left\{\frac{\Delta S_r^*}{R} + \ln\left(\frac{\kappa}{\gamma_{ABC^*}}\frac{k_B T}{h}\right)\right\} \tag{5.15}$$

For an ideal gas, the activation energy (E_a kJ/mol) is related to ΔH^*.

$$\Delta H^* = \Delta(E + PV) = \Delta E_a + \Delta n RT \tag{5.16}$$

In Eq. (5.2), 2 moles of reactants combine to form 1 mole of transition-state molecules, so $\Delta n = -1$ which makes $\Delta H^* = E_a - RT$. That means for an ideal gas at 25°C, ΔH^* is smaller than E_a by about 2.5 kJ/mol. This is a relatively small amount considering that E_a for most reactions are on the order of tens of kJ/mol. This energy difference can also be formulated in terms of the volume change associated with the formation of the activated complex. For an ideal gas, the volume change (ΔV^*) associated with the formation of a 1 molecule transition state from 2 molecules of reactants is approximately 2500×10^{-5} m³/mol. For aqueous solutions, which are featured in this chapter, the activation volume for reactions seldom exceeds 25×10^{-5} m³/mol (Drljaca et al., 1988), which is so much smaller than for the gas phase that the difference between E_a and ΔH^* for aqueous phase reactions is negligible. Because κ and γ are both approximately equal to one, they are typically ignored and the terms in brackets in Eq. (5.15) are combined into the pre-exponential constant, A.

$$\ln A = \left\{\frac{\Delta S_r^*}{R} + \ln\left(\frac{k_B T}{h}\right)\right\} \tag{5.17}$$

These adjustments result in the Arrhenius equation, which is used to correlate rate constants over a range of temperatures.

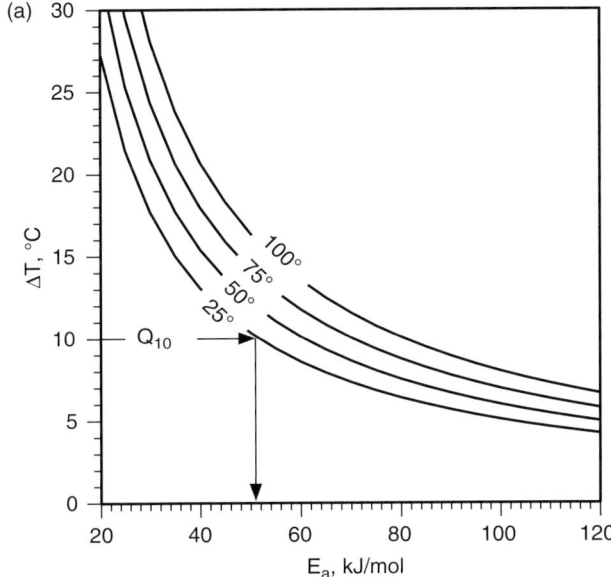

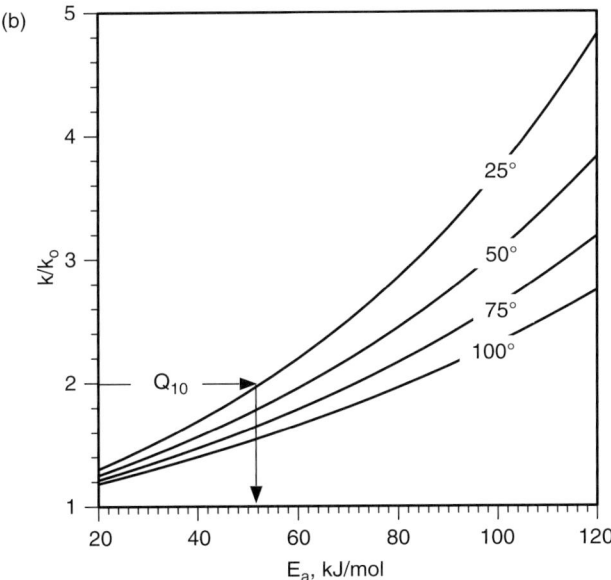

Figure 5.3. (a) The change in temperature that is needed to double the reaction rate as a function of the activation energy for the reaction. The contours show that ΔT changes as the temperature increases. (b) The relative change in the rate constant caused by a 10°C increase in temperature as a function of activation energy. The contours show that k/k_o decreases as the temperature increases.

$$\ln k = \left(\frac{-E_a}{R}\right)\frac{1}{T} + \ln A \qquad (5.18)$$

The Arrhenius equation is frequently recast to log base 10.

$$\log k = \left(\frac{-E_a}{2.303R}\right)\frac{1}{T} + \log A \qquad (5.19)$$

It can be converted into another convenient form by integrating its derivative between two temperatures.

$$\log\left(\frac{k_2}{k_1}\right) = \frac{-E_a}{2.303R}\left(\frac{1}{T_2} - \frac{1}{T_1}\right) \qquad (5.20)$$

This equation is the basis for the Q10 rule of thumb, which states that for many reactions the rate doubles for every 10°C increase in temperature. This approximation seems to work well for biochemical reactions near room temperature but is somewhat problematic for geochemical reactions that occur over a wider range of temperatures and have a wider range of activation energies. The effect of temperature and activation energy on doubling is shown in Figure 5.3a. Figure 5.3b shows how many times the rate will increase for every 10°C increase in temperature. These figures show that the Q10 rule applies for reactions that have an activation energy near 50 kJ/mol and occur near 25°C. Note that for a reaction with a low activation energy, the temperature must increase by a large amount to double the rate (Figure 5.3a); and for a reaction with a high activation energy only a small temperature increase will double the rate.

Example 5.2. Finding the activation energy for the ferric thiosulfate reaction using the Arrhenius equation

Williamson and Rimstidt (1993) determined the rate of the ferric thiosulfate reaction at temperatures ranging from 12° to 34°C. When those data are plotted on a graph of log k versus $1/T$ (Figure 5.4) the equation for the best-fit straight line is

$$\log k = 23.9 - 6426/T$$

The activation energy for this reaction can be calculated from the slope of this line.

$$E_a = 2.303R(6426) = 123{,}040 \text{ J/mol} = 123 \text{ kJ/mol}$$

The relatively large value of this activation energy is explained by the fact that the two positively charged ions must have enough energy to overcome their electrostatic repulsion. A postulated geometry for the activated complex is shown in Figure 5.5. Note that the reactants approach each other in a way that brings the negatively charged thiosulfate adjacent to the positively charged ferric iron ion of the other reactant molecule and vice versa. This geometry keeps the highly charged ferric iron ions separated as far as possible.

The very large activation energy of the ferric thiosulfate reaction means that a 5°C temperature increase, from 25° to ~30°C, will cause the rate to double (Figure 5.3).

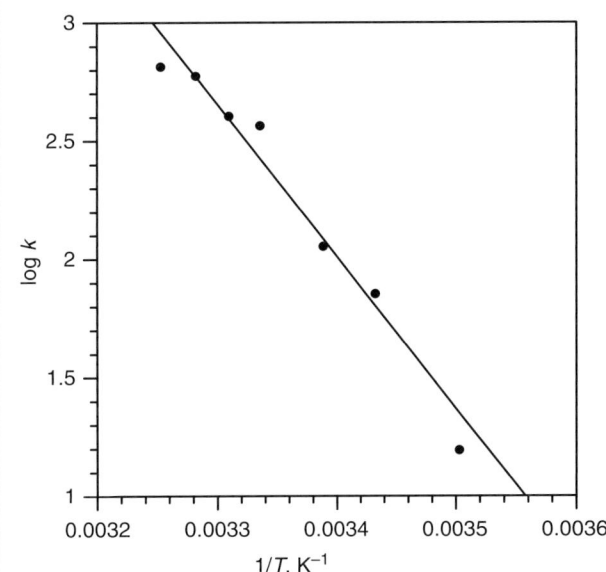

Figure 5.4. Arrhenius plot for the ferric thiosulfate reaction. The slope of the log k versus $1/T$ graph is 6426 so the activation energy for the reaction is 123 kJ/mol.

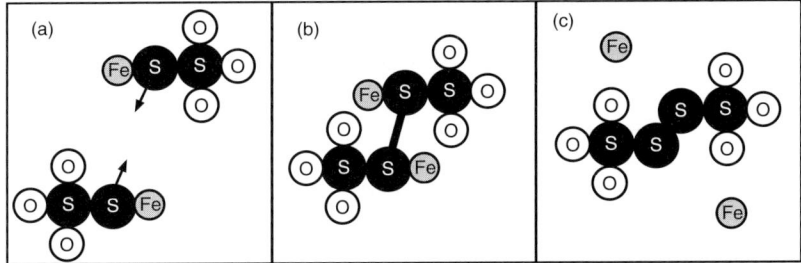

Figure 5.5. Schematic diagram showing the steps in the ferric thiosulfate reaction. (a) Two $FeS_2O_3^+$ molecules approaching each other. (b) The approximate geometry of the activated complex. (c) The reaction products, which are two ferrous iron ions and a tetrathionate ion.

Compensation effect

Several studies have reported that for a series of related reactions the pre-exponential in the Arrhenius equation correlates with the activation energy. This correlation has a physical significance for homogeneous, gas phase reactions but there is no theoretical justification for reactions involving condensed phases (Garn, 1975). Compensation effect models should be viewed with caution because errors in real data are known to cause a correlation between the slope and intercept of regression models (Liu and Guo, 2001;

Rimstidt *et al.*, 2012). Comparing the log k values at 298 K to the activation energy can minimize the correlation due to these errors (Ohlin *et al.*, 2010).

Effect of ionic strength on rates (Brönsted–Bjerrum equation)

The quasi-equilibrium model provides a simple way to account for the effect of ionic strength on the rates of aqueous reactions. In this model the ions A and B combine to form an activated complex that breaks down to products.

$$A + B = AB^* \to \text{products} \tag{5.21}$$

At infinite dilution ($I = 0$) where the activity coefficient for the activated complex equals one (and assuming that $\kappa = 1$) the rate constant for this reaction is related to the equilibrium constant for the formation of the activated complex.

$$k_o = \left(\frac{k_B T}{h}\right) K^* \tag{5.22}$$

If the rate equation is written in terms of concentration rather than activity, the activity coefficients are incorporated into the rate constant so the rate constant changes with changing ionic strength.

$$k = \left(\frac{\gamma_A \gamma_B}{\gamma_{AB^*}}\right)\left(\frac{k_B T}{h}\right) K^* = \left(\frac{\gamma_A \gamma_B}{\gamma_{AB^*}}\right) k_o \tag{5.23}$$

The Debye–Hückel limiting law relates the activity coefficients of all the species, including the activated complex, to the ionic strength.

$$-\log \gamma_i = A Z^2 \sqrt{I} \tag{5.24}$$

Z_i is the charge on the ion; I is the ionic strength of the solution; and $A = 0.5092$ (at 25°C). Equations (5.22) and (5.23) can be combined to create a relationship between log k and ionic strength.

$$\log k = \log k_o + \left(Z_A^2 + Z_B^2 - (Z_A + Z_B)^2\right) A \sqrt{I} \tag{5.25}$$

$Z_A + Z_B$ is the charge on the transition state. This equation can be simplified.

$$\log k = \log k_o + 2 Z_A Z_B A \sqrt{I} = \log k_o + 1.02 Z_A Z_B \sqrt{I} \tag{5.26}$$

This equation predicts that a graph of log k versus $\sqrt{I}$ will be a straight line with a slope of $2 A Z_A Z_B$. At 25°C, $2A = 1.02 \approx 1$, so the slope of the line is approximately equal to the product of the charges on the A and B ions. This relationship is not only an effective way to predict the effect of ionic strength on rates but it can also be used to confirm the stoichiometry of postulated transition states.

Example 5.3. Testing the postulated stoichiometry of the ferric thiosulfate transition state using rate versus ionic strength data

In the previous two examples, the activated complex for the ferric thiosulfate reaction was postulated to be $[(FeS_2O_3)_2^{2+}]^*$. This molecule is a combination of two $FeS_2O_3^+$ ions, each with a +1 charge, so the product $Z_A Z_B$ is +1. If this postulated stoichiometry for the activated complex is correct, the slope of a graph of log k versus $\sqrt{I}$ (Figure 5.6) should be +1. The Williamson and Rimstidt (1993) data for the rate as a function of ionic strength produces a straight line with a slope of 0.928.

$$\log k = 0.928\sqrt{I} + 2.17$$

Multiplying the slope of the line by 1.02 gives a value of 0.945 for the product of the charges on the reacting ions, which is very close to the expected value of +1. Note that this equation can be used to predict that log k_o = 2.17. This test gives further support to the postulated composition of the activated complex.

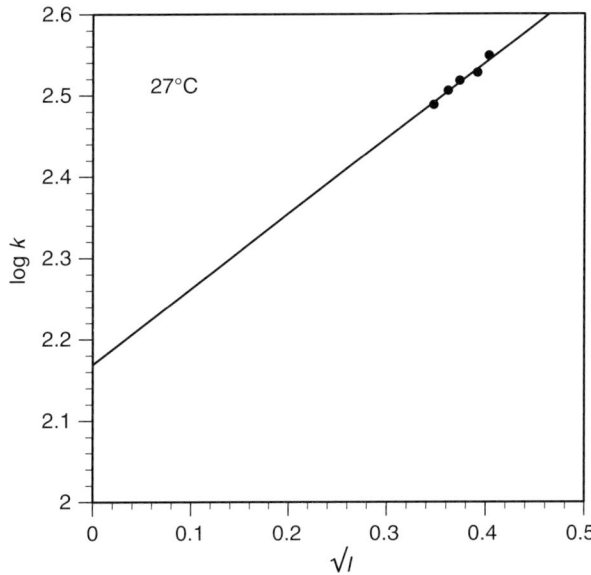

Figure 5.6. The graph of log k versus $\sqrt{I}$ for the ferric thiosulfate reaction. This graph has a slope of 0.928 ($\approx$ 1), which is consistent with each of the interacting molecules having a charge of +1.

Effect of pressure on rates

In addition to the effects of temperature and ionic strength, rates are also affected by changing pressure. By analogy to the effect of pressure on equilibrium constants, the expected effect of pressure on rate constants (Asano and le Nobel, 1978; Drljaca et al., 1988; Eckert, 1972) is related to the change in volume that results from the conversion of the reactants to the activated complex. This activation volume (ΔV^*) is the difference between the partial molal volumes of the reactants and the transition state.

$$\Delta V^* = \bar{V}_{AB^*} - \bar{V}_A - \bar{V}_B \tag{5.27}$$

van Dldik et al. (1989) have compiled a large amount of data pertaining to the effect of pressure on reactions in aqueous media. For the reaction, A + B = AB* → products, the rate constant change with pressure is proportional to the volume of activation.

$$\left(\frac{\partial \ln k}{\partial P}\right)_T = -\frac{\Delta V^*}{RT} \quad \text{or} \quad \left(\frac{\partial \log k}{\partial P}\right)_T = -\frac{\Delta V^*}{2.303 RT} \tag{5.28}$$

Two kinds of volume of change occur when dissolved species react to form an activated complex. The volume occupied by the species themselves changes and the associated water molecules undergo rearrangements that change their volume.

When Eq. (5.28) is integrated between 1 bar where the rate constant is k_1 and P where the rate constant is k_2, log (k_2/k_1) is predicted to be a linear function of pressure.

$$\log\left(\frac{k_2}{k_1}\right) = -\frac{\Delta V^*}{2.303 RT} P \tag{5.29}$$

Actual log k versus P relationships usually display some curvature over large ranges of pressure so the log k values are often fitted to a parabolic function of pressure.

$$\log k = a + bP + cP^2 \tag{5.30}$$

The activation volume is calculated from the slope of this fit at $P \approx 0$ (usually at 0.1 MPa)

$$\left(\frac{\partial \log k}{\partial P}\right)_T = b \tag{5.31}$$

A warning is appropriate here. If the rates are expressed using volume units such as molarity, the compressibility of the solvent must be considered when calculating the activation volume. This problem can be avoided by using pressure-independent units such as mole fraction or molality.

Example 5.4. The activation volume for Mn(II)-catalyzed SO_2 oxidation

Huss Jr. *et al.* (1982) measured the rate of Mn(II)-catalyzed oxidation of SO_2 by dissolved oxygen at pressures up to about 140 MPa (1400 bars). A fit of log R versus P for the data shown in Figure 5.7 gives the equation

$$\log k = 7.76 \times 10^{-7} P^2 - 1.95 \times 10^{-3} P - 5.47 \quad (5.32)$$

$$\frac{d \log k}{dP} = 2(7.76 \times 10^{-7})P - 1.95 \times 10^{-3} \quad (5.33)$$

The slope of this line when $P = 0$ is -1.95×10^{-3} MPa^{-1}.

$$\Delta V^* = -(2.303)(298.15\text{K})\left(8.314 \times 10^{-6} \frac{\text{m}^3\text{MPa}}{\text{K mol}}\right)\left(-1.95 \times 10^{-3} \frac{1}{\text{MPa}}\right)$$

$$\Delta V^* = 1.11 \times 10^{-5} \frac{\text{m}^3}{\text{mol}} \quad (5.34)$$

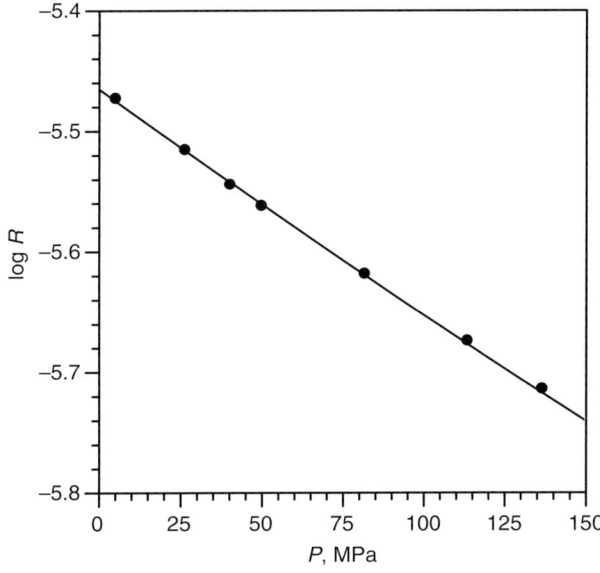

Figure 5.7. The effect of pressure on the Mn-catalyzed rate of oxidation of dissolved SO_2. The slope of the log R versus P graph is used to calculate the activation volume, ΔV^*, for the reaction.

Electron transfer reactions

Oxidation–reduction reactions, which transfer electrons from a donor species to an acceptor species, play a central role in many biological and geological processes. Unlike strongly coupled reactions where the reactants form a structurally defined activated complex, as described by transition-state theory, the species that participate in electron transfer reactions are weakly coupled and retain their individuality. R.A. Marcus (Marcus, 1964, 1968, 1985) developed the foundation of electron transfer theory.

Electron transfer reactions involve five steps (Astruc, 1995).

1. The donor and acceptor species come together to form a precursor complex.
2. Thermal activation allows solvent molecules in the precursor complex to attain an optimized arrangement for electron transfer to occur, i.e. an activated complex forms.
3. An electron transfers from the donor to the acceptor to form a successor complex.
4. The successor complex relaxes to a ground state.
5. The products separate by diffusion.

The overall rate of an electron transfer reaction is controlled by the rearrangement of the solvent molecules in the precursor complex. This can be appreciated by considering the self-exchange of an electron from a donor ferrous iron ion to an acceptor ferric iron ion. The ions in self-exchange reactions like this one can be distinguished using radioactive tracers (Silverman and Dodson, 1952).

$$Fe^*(H_2O)_6^{2+} + Fe(H_2O)_6^{3+} = Fe^*(H_2O)_6^{3+} + Fe(H_2O)_6^{2+} \quad (5.35)$$

For this reaction to occur, the water molecules coordinating the ferrous ion at an average distance of 2.21Å must move closer to the ion because the average distance of water molecules coordinating a ferric iron ion is 2.05Å. The electrostatic force that keeps each water molecule near the ion is constantly perturbed by thermal vibrations that cause each water molecule to move toward and away from the ion as if it was attached by a spring. According to Hook's law, the potential energy associated with each bond is a parabolic function of the vibrational displacement, x.

$$E_m = \frac{1}{2}kx^2 \quad (5.36)$$

Figure 5.8 shows the parabolic relationships between the energy (E) of the donor and acceptor species and the average displacement (x) of the solvent molecules from their most stable configuration. The intersection of these two parabolas is the energy barrier to the electron transfer step (i.e. free energy of activation, ΔG^*). The free energy of activation is a function of the

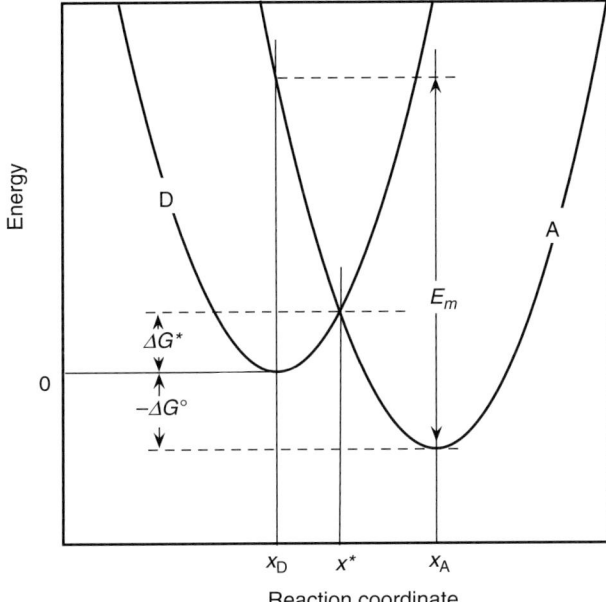

Figure 5.8. Graph of the energy of donor (D) and acceptor (A) species, and their waters of hydration, as a function of the displacement of the water molecules along the reaction coordinate (x). $\Delta G°$ is free energy driving the reaction, shown here as a negative value so the reaction is spontaneous. ΔG^* is the energy barrier that must be crossed for the electron to transfer from the donor to the acceptor. E_m is the reorganization energy.

reorganization energy (E_m or λ in many derivations, e.g. Marcus (2000)), and the standard state free energy of reaction ($\Delta G°$) (Houston, 2001).

$$\Delta G^* = \frac{(E_m + \Delta G°)^2}{4E_m} \quad (5.37)$$

The rate constant for the electron exchange reaction is the product of a frequency term and ΔG^*.

$$k = \frac{kT}{h} \exp\left(-\frac{(E_m + \Delta G°)^2}{4E_m kT}\right) \quad (5.38)$$

This simplified discussion of electron transfer for outer sphere interactions where the electron is transferred through solvent molecules is given to provide a conceptual basis for understanding this kind of reaction. Many electron transfer reactions are more complicated due to quantum mechanics effects (electron tunneling) and inner sphere interactions (Astruc, 1995). Basolo and Pearson (1967) give more details about electron transfer reactions.

Photochemical reactions

Solar energy delivers enormous amounts of energy to the Earth's surface, making photochemical reactions very important in geochemical and biological processes. The global average solar energy incident at the top of the atmosphere is 342 W/m². Some 102 W/m² is reflected back into space and between 65 and 98 W/m² is absorbed by the atmosphere (Arking, 1996). That leaves between 142 and 175 W/m² as the average solar energy flux at the Earth's surface, so that over a year sunlight delivers between 4.5 and 5.5 million kJ/m² of energy to the Earth's surface. Much of that radiant energy is converted to heat but a significant fraction drives photochemical reactions, the most important of which is photosynthesis. Many other photochemical reactions are important to geochemistry. For example, photo-oxidation of Fe^{2+} (Southworth, 1995) and Mn^{2+} (Nico et al., 2002) exert an important control on the redox chemistry of stream water. Most of those photochemical reactions occur in surface waters and in films of water on mineral surfaces.

Planck's law states that the amount of energy that each photon carries depends upon its frequency (v, sec^{-1}), multiplied by Planck's constant (6.63 × 10^{-34} J/sec) and the frequency depends on the speed of light (c, 2.99 × 10^8 m/sec) and the wavelength (λ, m).

$$E(\text{J/photon}) = hv = \frac{hc}{\lambda} \tag{5.39}$$

This means that light with shorter wavelengths delivers a larger amount of energy per photon. Blue green light (λ = 480 to 500 nm), the most intense sunlight at the Earth's surface, delivers 240 to 250 kJ per Einstein. An Einstein is 1 mole of photons (N_A = 6.02 × 10^{23}) regardless of wavelength.

$$E(\text{J/Einstein}) = \frac{N_A hc}{\lambda} \tag{5.40}$$

Figure 5.9 shows the relative intensity of sunlight and the amount of energy delivered by one Einstein of light as a function of wavelength.

When a molecule or ion absorbs a photon, that photon's energy can be dissipated in several different ways, but one way is for that energy to cause a chemical reaction to occur. The first law of photochemistry is that a compound must absorb light for a photochemical reaction to occur (Grotthuss–Draper law). The second law of photochemistry is that each photon that is absorbed activates only one molecule for a subsequent reaction (Stark–Einstein law). The quantum yield (ϕ) is defined as the number of molecules that react divided by the number of photons absorbed. It can also be defined in terms of moles.

$$\phi = \frac{\text{number of moles that photoreact}}{\text{number of Einsteins absorbed}} \tag{5.41}$$

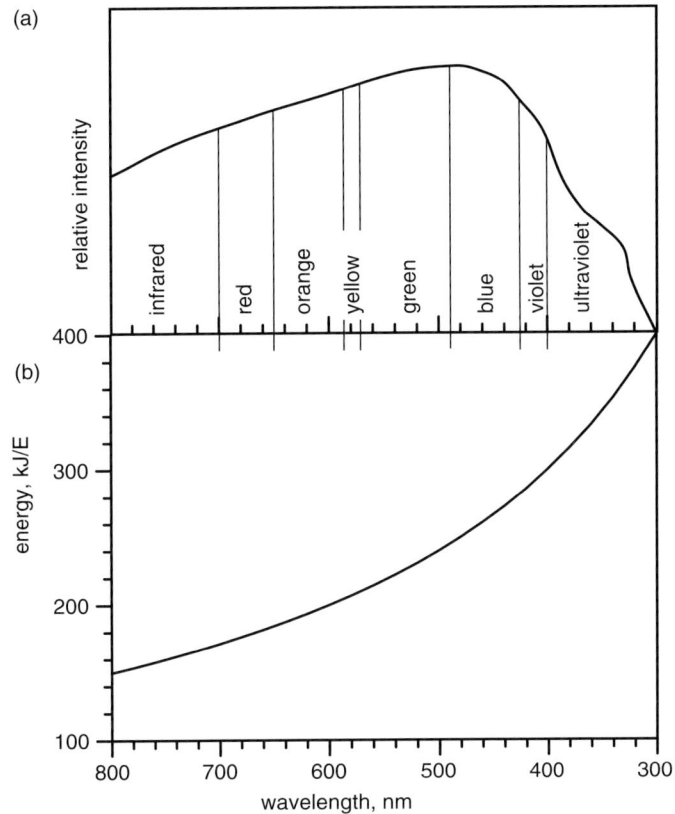

Figure 5.9. (a) Graph of the relative intensity of solar radiation at the Earth's surface. (b) Graph of the energy contained in 1 mole of photons as a function of wavelength of light.

Leifer (1988) provides useful theoretical and practical knowledge about aquatic photochemistry.

Radiolysis

Radiolysis reactions are brought about by the absorption of ionizing radiation. The radiation may belong to the light group, i.e. X-rays, γ-rays, electrons–positrons, muons; or the heavy group, i.e. protons, α-particles, fission fragments. When these particles pass through a medium, some of their energy is transferred to the molecules of that medium raising their electronic states. This can lead to neutral dissociation or to ionization if the energy transfer is sufficiently large.

Radiolysis of water produces hydroxyl free radicals (•OH) (Le Caër, 2011) along with hydrated electrons and •H atoms. Subsequent reactions produce H_2O_2, H_2, O_2, and, depending upon the solution composition, various other species. Because the H_2 can rapidly diffuse away from the site of generation, radiolysis reactions often alter the local redox conditions. For example, uranium deposits frequently show post-depositional oxidation because of radiolysis (Debessy et al., 1988; Derome et al., 2003; Smetannikov, 2011). Free radicals and hydrogen peroxide produced by the radiolysis of water in the subsurface can oxidize pyrite in the absence of O_2 (Lefticariu et al., 2010). Products of water radiolysis increase the leaching rate of basalt (Yoneawa et al., 1996). Radiolysis may have been the first important source of atmospheric O_2 (Draganic, 2005) and Lin et al. (2005) postulated that the natural radiolysis of water could produce enough H_2 to sustain microbial communities in the deep subsurface.

The chemical effect of radiolysis is controlled by a combination of absorbed dose and the radiation chemical yield value (G). Absorbed dose is measured in grays (1 Gy = 1 J/kg) and the G value indicates how many molecules of a species are produced per 100 ev (1 ev = 1.602×10^{-19} J) of absorbed energy. Ershov and Gordeev (2008) provide a model for the yield of H_2, H_2O_2, and O_2 from the radiolysis of water and aqueous solutions and Spinks and Woods (1990) give details about the connection between reaction kinetics and radiolysis processes.

Free radicals

Free radicals are chemical species that have one or more unpaired electrons. They can have a positive, negative, or zero charge. Although a few radicals are stable (e.g. O_2 and NO) or highly persistent, most are highly reactive. Thermally activated, electron transfer, photochemical, and radiolysis processes can all produce free radicals. Free radicals are common reaction intermediates in organic and biochemical reactions and they are no doubt important in many organic geochemical processes (Dominé et al., 2002).

Free radicals play important roles in many inorganic geochemical processes. For example, hydroxyl free radicals are produced during pyrite oxidation by O_2. Presumably they form as the result of a modified Fenton reaction (Rimstidt and Vaughan, 2003). Free radicals are further involved as various sulfoxy species are oxidized to sulfate (Druschel et al., 2003). Oxidation of Fe^{2+} is another example of a reaction that involves free radical intermediates.

Breaking Si–O bonds by fracturing quartz produces O_3–Si• and Si–O• free radicals that can subsequently react with water to produce •OH radicals (Narayanasamy and Kubicki, 2005). The •OH species is quite toxic to cells and may be responsible for the lung scarring that is symptomatic of silicosis. •OH radicals can also alter DNA and may be responsible for the development of lung cancer (Fubini, 1998).

Free radicals are studied by generating them using radiolysis. These studies have produced extensive tabulations of rate constants for free radical reaction rates (Bielski et al., 1985; Buxton et al., 1988; Buxton et al., 1995; Neta et al., 1988).

Reaction mechanisms

A reaction mechanism is a model that explains, elementary step by elementary step, how reactants become products. Reaction mechanism models also explain how the rate of each elementary step is influenced by the structure of its transition state. Because several different reaction paths may be possible for a particular reaction, there is no way to *a priori* determine a reaction mechanism. Transition states cannot be observed directly because they are present at very low concentrations and have a fleeting existence. This means that reaction mechanisms must be inferred using a combination of forward models based on fundamental physical principles and reverse models calibrated by the outcomes of experimental tests. Detailed studies of many kinds of reactions have generated some guiding principles for working out reaction mechanisms (Ašperger, 2003; Moore and Pearson, 1981). Casey (2001) and Casey and Swaddle (2003) give some guidelines that are specific to the dissolution of oxide and hydroxide compounds, which are common in geochemical environments.

Chemical bonding and reaction mechanisms

Chemical reactions occur when electrons move from the set of chemical bonds that comprise the reactants to form a new set of chemical bonds that comprise the products. This means that the nature of chemical bonding must be understood in order to construct models of reaction mechanisms.

Even though a rigorous definition of a chemical bond does not exist, the idea that electrostatic forces hold together the atoms in a molecule is well established. When atoms approach closely enough to form a bond, the valence electrons are attracted by the positive charge in the space between the nuclei and they accumulate between the atoms. The buildup of negative charge between the atoms attracts the atoms' nuclei causing them to become bonded. G.N. Lewis developed a simple model whereby a chemical bond is regarded as a pair of electrons, with opposite spins, residing between the bonded nuclei (Lewis, 1916). This simple idea has far-ranging implications (Jensen, 1980) and formed the foundation for many models of chemical structures developed throughout the twentieth century. Valence bond theory, which grew out of the Lewis model, is the basis of Linus Pauling's famous book *The Nature of the Chemical Bond* (Pauling, 1960). Valence bond theory assumes that electrons are highly localized in chemical bonds as well as in non-bonded sites as lone pairs. A contrasting model, molecular orbital theory, assumes that electrons are delocalized into molecular orbitals, which have geometries that are described by quantum mechanics (Gimarc, 1974). Both models are useful and complementary (Gillespie and Robinson, 2006). Quantum mechanics models of electron distribution in molecules now provide a powerful visualization of the localized bond and lone pair electron distributions in the form of the electron localization function (ELF). ELF maps for a number of important earth materials are shown in Gibbs *et al.* (2005).

Lewis acid–base theory (Jensen, 1980) is an outgrowth of the Lewis model of chemical bonds. A Lewis acid is a chemical species that can accept an electron pair. Lewis acids can be cations like Fe^{3+} or Cu^{2+} or they can be species with empty or partially empty valence orbitals such as CO_2 or SO_2. Lewis bases can donate an electron pair. Lewis bases are anions like OH^- or S^{2-} or they can be species with lone pairs such as H_2O or NH_3. The transfer of cations from a solid, such as szomolnokite ($FeSO_4 \cdot H_2O$), to form a hydrated ferrous ion in solution is a typical Lewis acid–base reaction.

$$FeSO_4 \cdot H_2O + 5\ H_2O = Fe(H_2O)_6^{2+} + SO_4^{2-}\ (aq)$$

In this reaction, ferrous iron, a Lewis acid, reacts with the lone pair electrons on water molecules, which are Lewis bases, to form a hydrated cation, a Lewis adduct. Although the sulfate ion also interacts with water molecules, this interaction is much weaker than the Lewis acid–base interaction of the ferrous iron. Many, perhaps all, chemical reactions can be viewed as Lewis acid–base interactions, making this theory very useful for developing models of reaction mechanisms.

Organic chemists refer to Lewis acids as electrophiles because they are attracted to electron-rich sites on other molecules, and they refer to Lewis bases as nucleophiles because they are attracted to electron-deficient molecular sites. Ingold (1969) developed a classification of reaction mechanisms based on this idea and this classification is the foundation for modeling the mechanisms of organic reactions (Bruckner, 2002; Grossman, 1999). Casey (2001) and Casey and Swaddle (2003) adapted some of these principles to apply to the dissolution of oxides. The electron-rich, and therefore nucleophilic, sites on molecules are nicely visualized using the electron localization function (Gibbs *et al.*, 2005).

Donor–acceptor theory (Gutmann, 1978) further extends the Lewis bonding theory to explain how bonds in reacting molecules lengthen (become weaker) or shorten (become stronger) as a result of the Lewis base (electron donor) and Lewis acid (electron acceptor) interactions. Gutmann proposed three rules for predicting bond length variations during donor–acceptor interactions:

1. As donor and acceptor atoms approach each other, the induced rearrangement of electrons in both the donor and the acceptor causes the adjacent bonds in each to lengthen.
2. Donor–acceptor interactions cause electron density shifts throughout the entire molecule and those shifts produce changes in bond lengths.
3. As the coordination number of an atom in an adduct increases so do the bond lengths of any bonds originating from the coordination center.

Chemical bonding and reaction mechanisms

Example 5.5. Hydrolysis of a Si–O bond

The donor–acceptor rules are consistent with quantum mechanical models that show how electrons in a bond respond as a donor molecule approaches the electrophilic site on an acceptor molecule. A well-known quantum mechanics model by Lasaga and Gibbs (1990) illustrates the attack of a H_2O molecule on a Si–O bond in a $H_6Si_2O_5$ molecule. Their model shows that the electron pair on the water molecule's oxygen atom mounts a nucleophilic attack on one of the Si atoms leading to a 5-coordinated Si atom. At the same time, one of the water molecule's hydrogen atoms mounts an electrophilic attack on the bridging oxygen in the $H_6Si_2O_5$ molecule. The simultaneous alignment of the oxygen end of the water molecule with the Si atom and the electrophilic attack of the hydrogen atom on the bridging oxygen atom creates the geometry of the transition state (Figure 5.10). Once that transition state has formed, the curly arrows show the shifts in electron density from preexisting bonds to the developing bonds. This shift divides the $H_6Si_2O_5$ molecule into two H_4SiO_4 molecules.

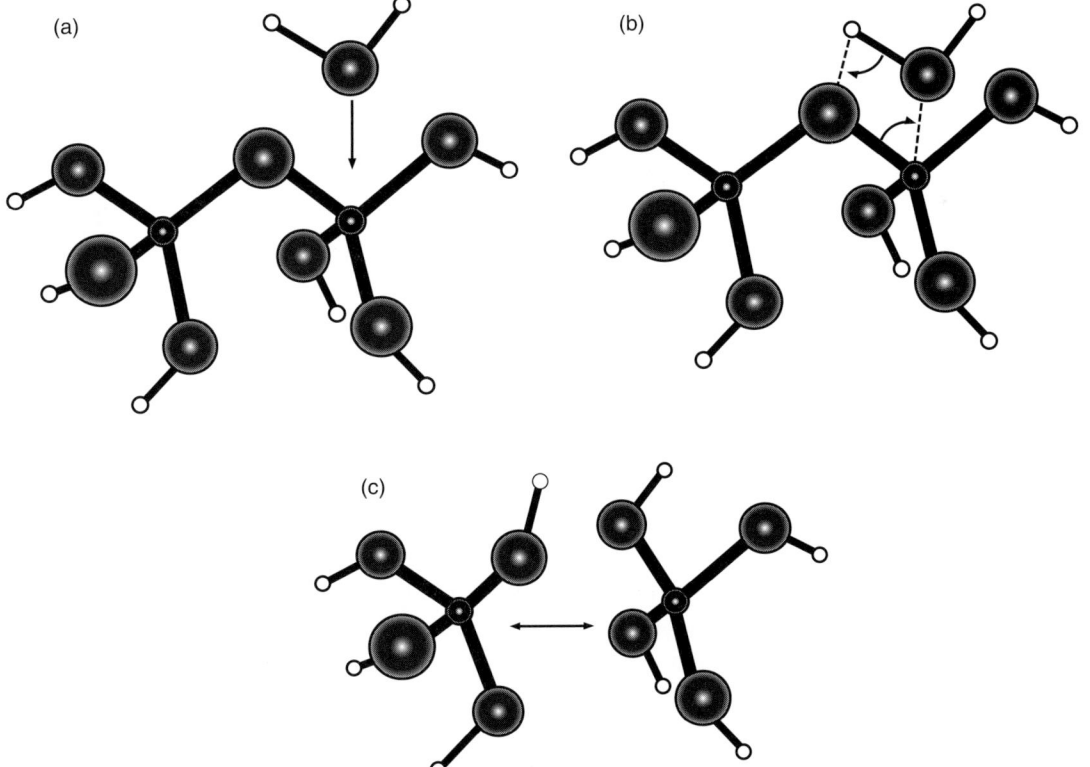

Figure 5.10. Schematic illustration showing how nucleophilic and electrophilic interactions cause (a) a water molecule to align with a Si–O bond to form (b) a transition state that results in electron rearrangement, which disassociates the Si–O bond and forms (c) two Si–OH moieties.

Rate equations and reaction mechanisms

A postulated reaction mechanism must be consistent with the rate data as well as the general principles of chemistry and physics described in this chapter. This congruency between the rate equation, general physicochemical principles, and the reaction mechanism is a foundation of chemical kinetics. Although many different guidelines might be formulated for creating this congruency, the ones listed below are often useful (Brezonik, 1994; Edwards *et al.*, 1968).

1. For elementary reactions, the composition of the transition state is reflected by the rate equation so that the number of molecules of each species found in the transition state is given by the reaction order for that species, i.e. the molecularity of the transition state is inferred from the reaction order for each species found in the rate equation. If the rate law is $r = km_A^a m_B^b$, we infer that the reaction is $aA + bB \rightarrow [A_aB_b]^* \rightarrow$ products.
2. Solvent molecules may participate in the activated complex but unless the concentration (activity) of the solvent can be varied widely this situation is difficult to recognize. Therefore, the activity of the solvent is usually set to one in the rate law. For the reaction $aA + bH_2O = cC$, the rate law would be $r = km_A^a a_{H_2O}^b = km_A^a$. Note that the composition of the solvent can affect the rate even if solvent molecules do not participate in the transition state (Amis, 1966).
3. Most reactions are bimolecular and only a few are either monomolecular or termolecular. When the overall reaction order is greater than three, it is likely that it is a chain reaction with one or more equilibrated intermediates prior to the rate-determining step. For the reactions $2A + 2B = 2AB$ (fast) and $2AB \rightarrow 2C$ (slow), the empirical rate equation might be written as $r = km_A^2 m_B^2$ and appear to be an overall fourth-order reaction but it would be more meaningful to write the rate law in terms of the intermediate AB, i.e. $r = km_{AB}^2$.
4. Rate equations that display non-integer (fractional) orders typically arise from chain (multistep) reactions, often involving free radical intermediates.
5. Rate equations that display non-integer (fractional) orders sometimes arise from rapid equilibria (sometimes competitive) prior to the rate-determining step.
6. The rate-determining step for the forward reaction sequence must be the same as the rate-determining step for the reverse reaction sequence. This is required by the principle of microscopic reversibility.
7. All intermediates produced in elementary steps must be consumed in subsequent steps otherwise they will show up as reaction products.
8. If the stoichiometric coefficient for species i is greater than the reaction order for i, one or more reaction steps occur after the rate-determining step.
9. If the reaction order increases as the concentration of one (or more) of the reactants increases, the overall reaction may consist of two or more parallel paths.
10. Highly reactive intermediates, which typically remain at a low concentration, are more likely to react with stable species, which occur at high concentrations, rather than reacting with each other.

11. Elementary reactions must be feasible from a bond energy and geometry standpoint. Extensive bond and atom rearrangement does not occur in a single elementary step.
12. The charge on the transition state of the rate-determining step can be deduced from the effect of ionic strength on the rate.

Catalysts and inhibitors

The previous section describes several strategies for interpreting experimental results to infer the nature of a reaction mechanism. Identifying species that catalyze or inhibit the reaction rate can provide additional valuable evidence about the reaction. A catalyst is a species that participates in the reaction and increases the reaction rate but emerges unchanged after the reaction is complete. For example, Dove (1994) has shown that the presence of Na^+ ions in solution increases the rate of quartz dissolution but the Na^+ is not consumed by the reaction. A catalyst lowers the activation energy for the reaction's rate-determining step. An inhibitor is a negative catalyst so it is a species that participates in a reaction and decreases its rate without being consumed by the reaction. Catalysts and inhibitors are very useful chemical probes that can be used to test hypothesized reaction mechanisms.

Example 5.6. Cu^{2+} catalysis of the ferric thiosulfate reaction

The model of the ferric thiosulfate reaction mechanism developed in Examples 5.3, 5.4, and 5.5 can be further tested using catalysts and inhibitors. The slow rate and high activation energy of this reaction is the result of the work required to bring the two positively charged $FeS_2O_3^+$ species together to form the transition state. This reaction would be much faster if the reacting species had zero charge.

Adding a small amount of Cu^{2+} to the reaction dramatically increases the rate. The Cu^{2+} forms a thiosulfate complex and because it is uncharged the $CuS_2O_3^0$ species can react more easily with a $FeS_2O_3^+$ so that the solution contains a higher concentration of activated complexes.

$$CuS_2O_3^0 + FeS_2O_3^+ = (CuS_2O_3 \cdot FeS_2O_3)^+ \rightarrow Cu^+ + Fe^{2+} + S_4O_6^2 \quad (5.42)$$

The Cu^+ is rapidly re-oxidized by a very fast reaction with ferric iron, which converts it back to Cu^{2+}.

$$Cu^+ + Fe^{3+} = Cu^{2+} + Fe^{2+} \quad (5.43)$$

This meets the definition of a catalyst, which is a species or site that lowers the activation energy of the reaction and is not consumed by the reaction.

The ferric thiosulfate reaction can be inhibited by the addition of Ag^+, which forms a very strong complex with thiosulfate and that complex is unreactive because Ag^+ cannot be reduced to Ag^0 by reaction with thiosulfate or oxidized by reaction with Fe^{3+}.

Chapter 6
Surface kinetics

Many important reactions happen at interfaces. Because reactions between aqueous species and mineral surfaces are so important in low temperature geochemistry, this chapter focuses on reactions at solid/solution interfaces.

Reaction rates at solid/solution interfaces are controlled by the area of the interface as well as by the chemical and physical conditions that occur there. Surface reactions are approximately confined to a two-dimensional region, so their rates are expressed in terms of how fast species are created per unit of surface area, and this means that the rates have units of flux (J, mol/m^{2}sec). The flux notation (J) and terminology is used throughout this book. The environment at the solid/solution interface is a hybrid of the bulk solid and bulk solution, so models of the chemical and physical conditions controlling the reaction rates must account for this transitional character. Equilibrium thermodynamics provides a powerful starting point for constraining the surface conditions. At equilibrium the chemical potential of each component must be the same throughout the system, so the chemical potential of the components in the surface are equal to their chemical potentials in the solid and solution phases. At low temperatures, the slow rate of equilibration between the bulk solid and the surface may void this requirement for the solid but it should apply for the components in the bulk solution. Also, at equilibrium the principle of detailed balance requires that the rates of forward reactions in the interface must equal the rates of the reverse reactions. In addition, the forward and reverse reaction steps must be the same. Models of reaction rates at equilibrium are well constrained by these principles but as the system departs from equilibrium these requirements fall away and we must search for other principles to model interfacial reaction rates.

Creating a new interfacial area requires work to be done and the amount of work needed to create a unit area of new surface is the surface free energy (σ, J/m^2).

$$\sigma = \left(\frac{dG}{dA}\right)_{T,P,\text{etc.}} \tag{6.1}$$

The surface free energy term is always positive; this means that the equilibrium surface should acquire a geometry that minimizes its surface free energy. An atomically smooth sphere has the lowest overall surface free energy for solids with isotropic bonding. For example, amorphous silica tends to develop a spherical form. For crystalline solids, the faces with the highest atomic packing densities and shortest bond lengths have the lowest σ values. These "low index" faces tend to occur more frequently in crystal forms, although the growing crystals are not at equilibrium with the surrounding solution so most crystals actually display growth forms rather than true equilibrium forms. A crystal that is at perfect equilibrium with the surrounding solution would have atomically smooth faces, but actual crystals usually display growth hummocks and other surface irregularities related to the growth process.

The value of σ for a crystal face relative to the other crystal faces can be found using Wulff's theorem, which states that there exists an interior point such that the surface free energy of a crystal face is proportional to the perpendicular distance from that point to the crystal face (Wolff and Gualtieri, 1962).

Geometric surface area models

Because reaction rates between minerals and solutions are directly proportional to the interfacial area between the phases, it is necessary to quantify this area for rate models. There are various methods to measure the interfacial area, but a useful first step for model building is to develop idealized reference models for reacting surfaces. Idealized surface area models neglect important surface features in a trade-off for simplicity. As such they provide a handy approximation of the relationship between surface geometry and reaction rates.

As a first approximation, the surface areas of solid particles can be modeled as if the grains are smooth spheres with an effective diameter equal to the real grain's smallest dimension. There are several methods to determine particle size, including sieving, sedimentation rates, light scattering, and optical measurements. All these methods will find a range of grain sizes. If this range is large, the surface area models must account for the particle size distribution. If the range is small, a weighted average of the maximum diameter (D_{max}, m) and minimum diameter (D_{min}, m) is used to determine an effective diameter (D_e, m). The averaging method of Tester et al. (1994), which assumes a flat particle size distribution between the maximum and minimum diameters, seems to work well.

$$D_e = \frac{D_{max} - D_{min}}{\ln\left(\frac{D_{max}}{D_{min}}\right)} \tag{6.2}$$

The effective diameter given by this model is very close to the arithmetic average of the maximum and minimum grain diameter for a narrow grain size range, but for a wider size range it is lower because the smaller particles contribute more surface area.

The simplest model that relates the specific surface area of grains to their diameter assumes that the grains are spherical. This model expresses the specific geometric surface area (A_{geo}, m²/g) as a function of the molar volume (V_m, m³/mol) of the substance, the diameter of the grains (D_e, m), and the molecular weight of the substance (W_m, g/mol).

$$A_{geo} = \frac{6V_m}{DW_m} \left(\frac{m^2}{g}\right) \tag{6.3}$$

Equation (6.3) can be transformed to give the specific surface area as a function of the density (ρ, g/cm³) and diameter (D, m) of the grains.

$$A_{geo} = \frac{6 \times 10^{-6}}{D\rho} \left(\frac{m^2}{g}\right) \tag{6.4}$$

Example 6.1. Geometric specific surface area of quartz grains

If a sample of quartz sand is sieved to recover the 1 to 2 mm diameter size fraction, the effective diameter of the grains in the sample can be found using Eq. (6.2).

$$D_e = \frac{2-1}{\ln\left(\frac{2}{1}\right)} = 1.44 \text{ mm} = 1.44 \times 10^{-3} \text{ m}$$

The density of quartz is 2.66 g/cm³, so the geometric specific surface area of the sample is found using Eq. (6.4).

$$A_{geo} = \frac{6 \times 10^{-6}}{(1.44 \times 10^{-3})(2.66)} = 0.00157 \left(\frac{m^2}{g}\right)$$

Surface site density

Each surface reaction takes place at a discrete site where an attacking reagent docks and participates in an elementary reaction. The exact nature of the site varies from case to case but the overall reaction rate is proportional to the number of docking sites. If these sites are nearly homogeneously distributed over the surface, the number of reaction sites, and therefore the overall reaction rate, is directly proportional to the surface area. The number of *potential* sites per unit area is approximately equal to the number of formula units that crop out on the surface. The area occupied by one formula unit (A_{fu}, m²/fu) is a function of the molar volume (m³/mol) of the substance and Avogadro's number (N_A).

$$A_{fu} = \left(\frac{V_m}{N_A}\right)^{2/3} \quad (6.5)$$

The area occupied by 1 mole of formula units (A_{mol}, m²/mol) is the area per formula unit multiplied by Avogadro's number.

$$A_{mol} = A_{fu} N_A \quad (6.6)$$

The formula unit density (D_{fu}, mol/m²) is the reciprocal of A_{mol}.

$$D_{fu} = \frac{1}{A_{mol}} \quad (6.7)$$

For a surface that is dissolving homogeneously and congruently, the average lifetime of a single formula unit (t_{fu}, sec) is the reciprocal of the dissolution flux (J, mol/m²sec) and the area occupied by 1 mole of formula units.

$$t_{fu} = \frac{1}{JA_{mol}} \quad (6.8)$$

Example 6.2. Average lifetime of a Mg_2SiO_4 unit on a dissolving forsterite surface

Forsterite (Mg_2SiO_4) has a molar volume of 4.365×10^{-5} m³/mol. The average area occupied by one formula unit of forsterite is found using Eq. (6.5).

$$A_{fu} = \left(\frac{4.365 \times 10^{-5} \frac{m^3}{mol}}{6.023 \times 10^{23} \frac{fu}{mol}}\right)^{2/3} = 1.738 \times 10^{-19} \frac{m^2}{fu} \quad (6.9)$$

The surface area occupied by 1 mole of forsterite is found using Eq. (6.6).

$$A_{mol} = \left(1.738 \times 10^{-19} \frac{m^2}{fu}\right)\left(6.023 \times 10^{23} \frac{fu}{mol}\right) = 1.047 \times 10^5 \frac{m^2}{mol} \quad (6.10)$$

The site density is found using Eq. (6.7).

$$D_{fu} = \frac{1}{1.047 \times 10^5} = 9.553 \times 10^{-6} \frac{\text{mol}}{\text{m}^2} \tag{6.11}$$

The dissolution rate of forsterite at 25°C and pH 3 is 1.99×10^{-8} mol/m²sec so the average lifetime of a surface formula unit is found using Eq. (6.8).

$$t_{fu} = \frac{1}{\left(1.99 \times 10^{-8} \frac{\text{mol}}{\text{m}^2\text{sec}}\right)\left(1.047 \times 10^5 \frac{\text{m}^2}{\text{mol}}\right)} = 480 \text{ sec} \approx 8 \text{ min} \tag{6.12}$$

Surface area/volume models

Rate models that express the rate of change of concentration (R, molal/sec) due to the dissolution or precipitation of a solid must account for the surface area to volume (A/V) or surface area to mass of solution (A/M) ratio of the reacting system. Table 6.1 gives A/V and A/M ratios for some common geometries.

Example 6.3. *A/M ratio for packed beds*

The porosity of packed beds ranges from about 50% for loosely packed grains to about 25% for close packed grains. The A/M for a packed bed of 1 mm diameter grains with 30% porosity filled with a solution with a density of 1 g/cm³ is about 14 m²/kg.

$$A/M = \left(\frac{1-\phi}{\phi}\right)\left(\frac{6}{1000\rho D}\right) = \left(\frac{0.7}{0.3}\right)\left(\frac{6}{(1000)(0.001)}\right) = 14 \frac{\text{m}^2}{\text{kg}}$$

Figure 6.1 shows how the A/M ratios for packed beds filled with a solution with a density near 1 g/cm³ varies with grain diameter and porosity. Real grains are not perfect spheres so these values should be increased by 5 to 10 times to account for their surface roughness.

Reactions at surfaces

Surface reactions comprise four steps.

1. Adsorption of a fluid phase species (A) to a surface site (S): $A(aq \text{ or } v) + S = A \cdots S$
2. Surface diffusion of the reacting species to a reactive site (S#): $A \cdots S \rightarrow A \cdots S\#$
3. Reaction between the reactive site and the adsorbed species to produce a product species (B): $A \cdots S\# = A-S\# \rightarrow B-S$

Table 6.1. *A/V and A/M ratios for some typical solid/solution geometries. The A/V ratios are used for models where the solution concentrations are expressed in volume units. The A/M ratios are used in models where the solution concentrations are expressed in mass units. The units for density, ρ, are g/cm³.*

Geometry	A/V (m²/m³) and A/M (m²/kg) ratio
Cylindrical pipe radius (R, m)	$A/V = 2R$ and $A/M = \dfrac{2}{1000\rho R}$
Rectangular pipe height (H, m), width (W, m)	$A/V = \dfrac{2(H+W)}{HW}$ and $A/M = \dfrac{2(H+W)}{1000\rho HW}$
Fracture with parallel walls width (W, m)	$A/V = \dfrac{2}{W}$ and $A/M = \dfrac{2}{1000\rho W}$
Packed bed of spheres diameter (D, m), porosity (ϕ)	$A/V = \left(\dfrac{1-\phi}{\phi}\right)\left(\dfrac{6}{D}\right)$ and $A/M = \left(\dfrac{1-\phi}{\phi}\right)\left(\dfrac{6}{1000\rho D}\right)$

4. Desorption of the product species: B–S = B⋯S → B(*aq* or *v*) + S

Any of these steps can be rate limiting.

Adsorption (Langmuir model)

The adsorption step is often modeled using the Langmuir isotherm, which is developed from two primary assumptions: (1) only monolayer adsorption occurs and (2) the sorption enthalpy is the same for all sites regardless of the amount of surface coverage. This model was originally derived for adsorption of a gas onto a solid surface but it is developed here to model adsorption of a solution species to a solid surface. For the sorption process, the rate of change of fractional surface coverage ($\theta = \Gamma_a/\Gamma$, no units), which is the ratio of the surface concentration of occupied sites (Γ_a, mol/m²) to the surface concentration of all adsorption sites (Γ, mol/m²), is directly proportional to the adsorption rate constant (k_+, molal/sec), the concentration of adsorbate (m, molal), the number of surface sites, and the proportion of the surface that is not covered by adsorbate ($1 - \theta$). The rate of desorption is directly proportional to the desorption rate constant (k_-, sec^{-1}), the number of surface sites, and the fraction of those sites that are covered by adsorbate. The overall rate is the difference between the adsorption rate and the desorption rate.

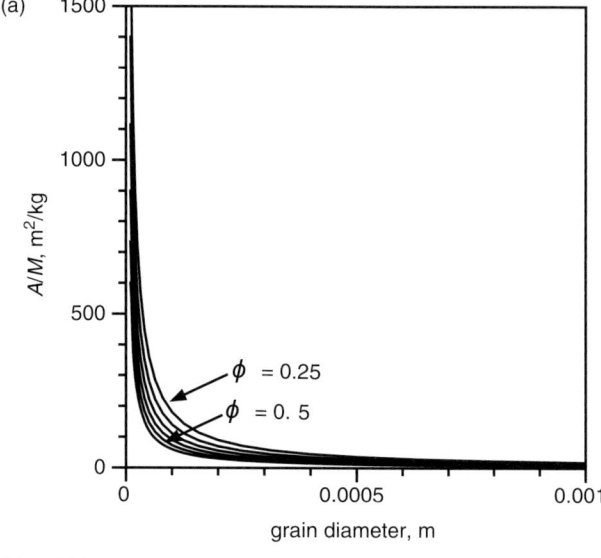

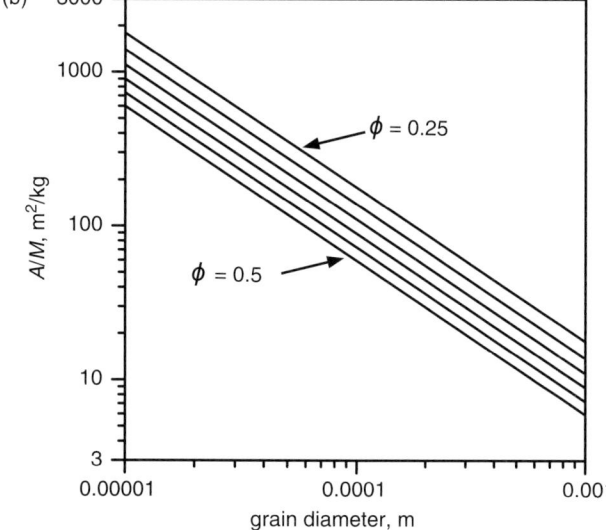

Figure 6.1. (a) A/M ratio for packed beds filled with a solution with a density of 1.0 g/cm³ as a function of grain diameter and bed porosity. (b) Same data presented on a graph with logarithmic scales.

$$\frac{d\theta}{dt} = k_+ m\Gamma(1-\theta) - k_-\Gamma\theta \qquad (6.13)$$

At equilibrium, the principle of detailed balance requires that the adsorption rate equals the desorption rate ($d\theta/dt = 0$).

$$k_+ m\Gamma(1-\theta) = k_-\Gamma\theta \qquad (6.14)$$

Equation (6.14) can be rearranged to define an apparent equilibrium constant (K_b, molal) for the sorption reaction.

$$\frac{k_+}{k_-} m \frac{\Gamma}{\Gamma} = \frac{\theta}{1-\theta} \qquad (6.15)$$

$$K_b m = \left(\frac{\theta}{1-\theta}\right) \qquad (6.16)$$

The fractional surface coverage is found by solving Eq. (6.16) for θ.

$$\theta = \frac{K_b m}{1 + K_b m} \qquad (6.17)$$

If θ is decomposed into Γ_a and Γ, Eq. (6.17) can be written as an isotherm that expresses surface concentration in terms of solution concentration.

$$\Gamma_a = \Gamma \frac{K_b m}{1 + K_b m} \qquad (6.18)$$

Figure 6.2 illustrates the relationship between the surface concentration and the solution concentration for some values of K_b.

Isotherm experiments, which determine Γ_a for a wide range of m, can be analyzed to find K_b and Γ_t. Equation (6.18) can be linearized in several different ways to avoid the need for using a nonlinear regression for data analysis. The double reciprocal of the Langmuir equation produces the Lineweaver–Burke equation, which is often used to fit adsorption data.

$$\frac{1}{\Gamma_a} = \frac{1}{\Gamma} + \frac{1}{\Gamma K_b m} \qquad (6.19)$$

A graph of $1/\Gamma_a$ versus $1/m$ has a slope of $1/\Gamma K_b$ and an intercept of $1/\Gamma$. This fit gives excessive weight to data in the low concentration range and is sensitive to data error. The Scatchard equation gives more weight to data in the high concentration range.

$$\frac{\Gamma_a}{m} = \Gamma K_b + K_b \Gamma_a \qquad (6.20)$$

According to this equation, a graph of Γ_a/m versus Γ_a has a slope of K_b and an intercept of ΓK_b. Another linear form proposed by Langmuir is less sensitive to data error and gives more weight to the middle and high concentration ranges.

$$\frac{m}{\Gamma_a} = \frac{m}{\Gamma} + \frac{1}{\Gamma K_b} \qquad (6.21)$$

A graph of m/Γ_a versus m has a slope of $1/\Gamma$ and an intercept of $1/\Gamma K_b$.

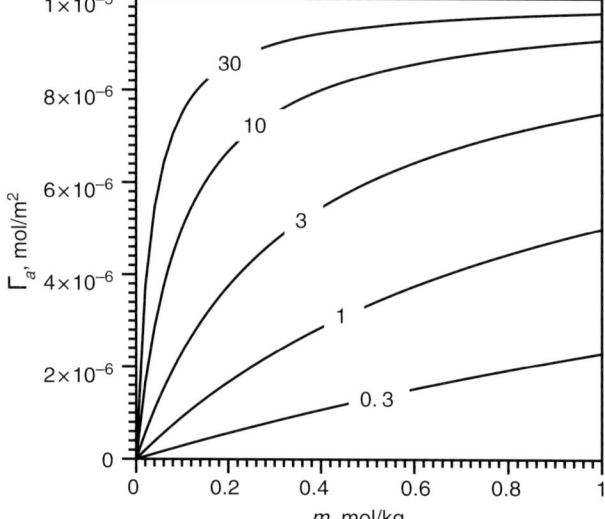

Figure 6.2. Langmuir sorption isotherms contoured for values of K_b ranging from 0.3 to 30. Γ is 1×10^{-5} mol/m² for each isotherm.

Distribution coefficient

The distribution coefficient is a simplification of the Langmuir equation, Eq. (6.16). For small amounts of surface coverage, the $1-\theta$ term in that equation is approximately equal to 1, so it simplifies to a linear relationship between the fractional surface coverage and the solution concentration.

$$K_b m = \theta \quad (6.22)$$

This equation can be rearranged and combined with the definition of fractional surface coverage ($\theta = \Gamma_a/\Gamma$) to find the surface concentration (Γ_a, mol/m²).

$$\Gamma_a = \Gamma\theta = \Gamma K_b m \quad (6.23)$$

The specific amount adsorbed (n_{sp}, mol/g-solid) is the surface concentration (Γ_a, mol/m²) multiplied by the specific surface area of the solid (A_{sp}, m²/g-solid).

$$n_{sp} = K_b m \Gamma A_{sp} \quad (6.24)$$

The distribution coefficient (K_D, (mol/g-solid)/molal) is the specific amount sorbed divided by the solution concentration.

$$K_D = \frac{n_{sp}}{m} \quad (6.25)$$

$$K_D = \frac{K_b m \Gamma A_{sp}}{m} = K_b \Gamma A_{sp} \qquad (6.26)$$

In this derivation, Γ_a and Γ_t are multiplied by the specific surface area of the solid in order to convert the concentration units from mol/m² to mol/g-solid. Defining distribution coefficients this way makes them functions of the specific surface area so K_D values vary with grain size.

BET surface area

Reaction rates between fluid species and solid surfaces are often normalized using a surface area value determined by the BET (Brunauer–Emmett–Teller) method. The theory behind this method was developed by Brunauer *et al.* (1938) and is thoroughly described and evaluated in Lowell and Shields (1991). This model is based on the idea that as the pressure and temperature of a gas approaches the pressure and temperature where liquid and vapor are in equilibrium, the molecules will deposit on the surface as a thin, multilayer film. In most cases N_2 is used as the gas and a powdered sample is placed into an evacuated chamber that is immersed into boiling liquid nitrogen to fix the temperature. Increasing amounts of N_2 gas are introduced into the chamber and the amount adsorbed at pressures ranging from 5 to 35 kPa is measured. Usually three to five data points are determined. These data are fit to the BET isotherm and the amount of gas that comprises a monolayer on the solid surface is computed from the slope and intercept of the isotherm. The surface area is then calculated assuming a sorption cross-section for an adsorbed N_2 molecule. Prior to the BET surface area determination, the surface of the sample is usually "cleaned" by heating in a vacuum, often to temperatures near 300°C. This treatment has little effect on most minerals but can drastically alter hydrous phases and sulfide minerals that decompose releasing water or sulfur vapor into the vacuum system. BET-determined surface areas are usually five to ten times larger than surface areas calculated using Eq. (6.3). For example, White and Peterson (1990) report that the average BET surface area for a relatively large number of samples is about seven times the surface area estimated from the geometric surface area estimated using Eq. (6.4).

Surface catalysis: unimolecular decomposition

Surfaces often catalyze chemical reactions. The simplest model of surface catalysis of a decomposition reaction involves two steps. First the reactant adsorbs to a surface to form a surface complex and then the surface complex decomposes to products.

$$A + S \underset{k_{-1}}{\overset{k_{+1}}{\rightleftharpoons}} AS \xrightarrow{k_{+2}} \text{products} \tag{6.27}$$

After a short time, the rate that A adsorbs to the surface equals the rate that AS breaks down to products. The rate of disappearance of A can be expressed in terms of the rate of conversion of AS to products.

$$\frac{dm_A}{dt} = k_{+2}\Gamma_{AS} = k_{+2}\theta\Gamma \tag{6.28}$$

Once the reaction reaches steady state, the rate of change of the AS concentration is zero.

$$\frac{d\Gamma_{AS}}{dt} = 0 = k_{+1}A\Gamma(1-\theta) - k_{-1}\Gamma\theta - k_{+2}\Gamma \tag{6.29}$$

This equation can be rearranged to find the fraction of surface sites covered by adsorbed A.

$$\theta = \frac{k_{+1}m_A}{k_{+1}m_A + k_{-1} + k_{+2}} \tag{6.30}$$

Equation (6.30) can be combined with Eq. (6.28) to find the rate of consumption of A.

$$\frac{dm_A}{dt} = \frac{k_{+1}k_{+2}\Gamma m_A}{k_{+1}m_A + k_{-1} + k_{+2}} \tag{6.31}$$

This result is analogous to the Michaelis–Menton equation that is used to describe enzyme-catalyzed decomposition rates.

If $k_{+2} \gg k_{+1}m_A$ and $k_{+2} \gg k_{-1}$, the rate-limiting step is adsorption.

$$\frac{dm_A}{dt} = \frac{k_{+1}k_{+2}\Gamma m_A}{\text{small} + \text{small} + k_{+2}} = k_{+1}\Gamma m_A \tag{6.32}$$

If $k_{+2} \ll k_{+1}m_A$ and $k_{+2} \ll k_{-1}$, the fraction of surface coverage is described by the Langmuir isotherm.

$$\theta = \frac{k_{+1}m_A}{k_{+1}m_A + k_{-1}} \tag{6.33}$$

This means that the rate depends upon the concentration of AS.

$$\frac{dm_A}{dt} = \frac{K_b k_{+2}\Gamma m_A}{K_b m_A + 1} \tag{6.34}$$

When m_A is small, so the fraction of surface coverage is small, the rate is first order in A.

$$\frac{dm_A}{dt} = K_b k_{+2} \Gamma m_A \qquad (6.35)$$

When m_A is large, so the fraction of surface coverage is large, the rate is zeroth order in A.

$$\frac{dm_A}{dt} = k_{+2} \Gamma \qquad (6.36)$$

Example 6.4. Oxidation rate of UO_2

In a detailed study of the dissolution rate of synthetic UO_2, de Pablo *et al.* (1999) found that instead of being linear, as is typical of many mineral dissolution reactions, the log J versus log M_{HCO_3} graph shows a distinct curvature. They interpreted this to mean that bicarbonate ions sorb to a surface site as a dissolution reaction step and when bicarbonate concentrations are high these sites become nearly saturated with adsorbed bicarbonate. This means that the reaction order transitions from first order at low bicarbonate concentration to zeroth order at high bicarbonate concentration. To model this behavior, they fit the rate data to

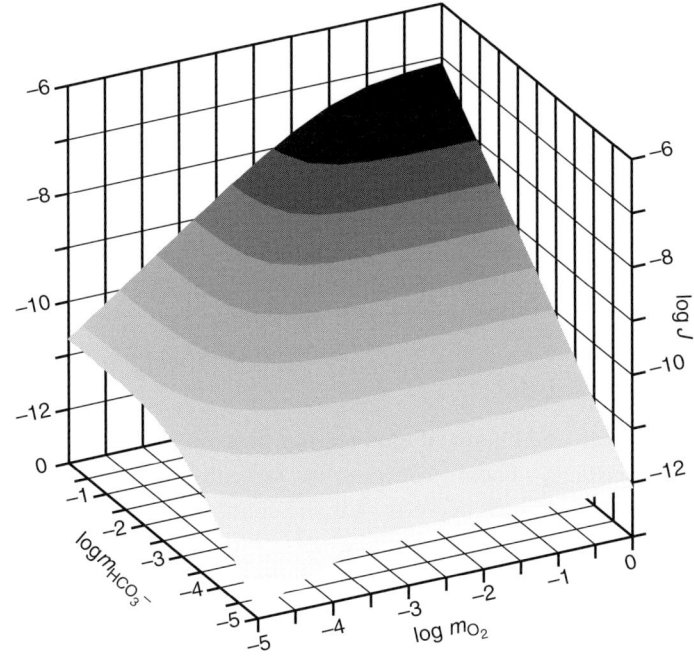

Figure 6.3. Graph of the uranium dissolution flux from oxidizing UO_2 as a function of bicarbonate and dissolved oxygen concentrations at 25°C, based on data from de Pablo *et al.* (1999).

an equation that assumes that both bicarbonate and oxygen interact with the dissolving surface via Langmuir isotherms.

$$J = \frac{k_{+1}k_{+2}\Gamma_{UO_2} m_{O_2} m_{HCO_3^-}}{k_{-1} + k_{+2}m_{HCO_3^-} + k_{+1}m_{O_2}} \tag{6.37}$$

A graph of log J as a function of log M_{HCO_3} and log M_{O_2} (Figure 6.3) shows that for high values of log M_{O_2} there is a linear relationship between log J and log M_{HCO_3}, but at low values of log M_{O_2} and high values log M_{HCO_3}, the log J values become independent of log M_{HCO_3}.

Surface retreat rate

For a dissolving flat surface, the linear rate of surface retreat (dz/dt, m/sec) is the dissolution flux (J, mol/m²sec) multiplied by the molar volume of the solid (V_m, m³/mol).

$$\frac{dz}{dt} = -JV_m \tag{6.38}$$

Example 6.5. Surface retreat rate for dissolving forsterite

The rate of forsterite dissolution in Example 6.2 is 1.99×10^{-8} mol/m²sec and the molar volume of forsterite is 4.365×10^{-5} m³/mol. At pH 3 and 25°C the surface retreat rate is calculated using Eq. (6.38).

$$\frac{dz}{dt} = -\left(1.99 \times 10^{-8} \frac{mol}{m^2 sec}\right)\left(4.365 \times 10^{-5} \frac{m^3}{mol}\right) = -8.69 \times 10^{-13} \frac{m}{sec} \tag{6.39}$$

The dissolving surface retreats at a rate of 27.4 µm/yr.

Shrinking particle model

Many hydrometallurgy and chemical engineering processes involve the dissolution of solid particles. These processes are typically modeled using a shrinking particle model, which accounts for the change in dissolution rate due to decreasing grain surface area as they dissolve away (Burkin, 2001; Levenspiel, 1972a). The classical derivation of the shrinking particle rate equation given here shows how the dissolution rate constant, k_+(mol/m²sec), can be calculated from the particle rate constant (k_p) that is derived from a fit of the extent of reaction versus time. The shrinking particle model assumes that the dissolution flux ($J = k_+$, mol/m²sec) is constant over the full extent of the reaction, either because the solution composition is constant, making the reaction pseudo-zeroth order, or because the reaction is actually zeroth order. The first step is to express the rate of reaction of a spherical particle (r, mol/sec) in terms of the particle's radius (R, m).

$$r = -Ak_+ = -4\pi R^2 k_+ \tag{6.40}$$

The rate of volume loss for the particle (dV/dt, m³/sec) is the rate of reaction multiplied by the molar volume (V_m, m³/mol).

$$\frac{dV}{dt} = rV_m = -4\pi R^2 k_+ V_m \tag{6.41}$$

The rate of volume loss of the particle is also given by the time derivative of the volume formula for a sphere.

$$\frac{dV}{dt} = 4\pi R^2 \frac{dR}{dt} \tag{6.42}$$

Setting Eq. (6.41) equal to Eq. (6.42) gives the rate of change of the radius.

$$\frac{dR}{dt} = -k_+ V_m \tag{6.43}$$

The fraction of material that is reacted away (α) at any time can be expressed in terms of the radius at that time (R, m) and the initial radius (R_o, m).

$$\alpha = 1 - \left(\frac{R^3}{R_o^3}\right) \tag{6.44}$$

$$R = R_o(1-\alpha)^{1/3} \tag{6.45}$$

The time derivative of Eq. (6.45) gives the rate of change of the fraction reacted in terms of the rate of change in the particle radius.

$$\frac{d\alpha}{dt} = -3\left(\frac{R^2}{R_o^3}\right)\frac{dR}{dt} \tag{6.46}$$

The right-hand side of this equation can be recast in terms of α by substituting Eq. (6.43) for dR/dt and Eq. (6.45) for R.

$$\frac{d\alpha}{dt} = \frac{3k_+ V_m}{R_o}(1-\alpha)^{2/3} \tag{6.47}$$

It is convenient to gather the constants on the right-hand side of this equation into a particle rate constant (k_p, sec^{-1}).

$$k_p = \frac{k_+ V_m}{R_o} \qquad (6.48)$$

This makes (6.47) into a simple function of α.

$$\frac{d\alpha}{dt} = 3k_p (1-\alpha)^{2/3} \qquad (6.49)$$

Integrating Eq. (6.49) and evaluating the integral gives a relationship between the fraction reacted and time.

$$\int_o^\alpha \frac{d\alpha}{(1-\alpha)^{2/3}} = 3k_p \int_0^t dt \qquad (6.50)$$

$$1 - (1-\alpha)^{1/3} = k_p t \qquad (6.51)$$

A graph of $1-(1-\alpha)^{1/3}$ versus t has the slope of k_p. The rate constant (k_+) for the dissolution flux is found by rearranging Eq. (6.48).

$$k_+ = \frac{R_o k_p}{V_m} \qquad (6.52)$$

Example 6.6. Forsterite dissolution flux determined using the shrinking particle model

Van Herk *et al.* (1989) proposed using forsterite (Mg_2SiO_4) to neutralize waste industrial acids. In order to provide rate data to support their idea, they performed a series of laboratory tests that determined the fraction of forsterite reacted away (α) by contact with acidic solutions for various times. Their data can be used, along with Eqs (6.51) and (6.52), to find the rate constant for forsterite dissolution.

The experiment dissolved 105–125 µm forsterite grains in pH 1 HCl solution at 40°C. The results are shown in Figure 6.4. When $1 - (1-\alpha)^{1/3}$ is graphed versus t, the slope of the resulting straight line is 1.23×10^{-6} sec^{-1}. The effective diameter of the particles can be estimated using Eq. (6.2).

$$D_e = \frac{D_{max} - D_{min}}{\ln\left(\frac{D_{max}}{D_{min}}\right)} = \frac{125 \times 10^{-6} - 105 \times 10^{-6}}{\ln\left(\frac{125 \times 10^{-6}}{105 \times 10^{-6}}\right)} = \frac{20 \times 10^{-6}}{0.174} = 115 \times 10^{-6} \text{ m} \qquad (6.53)$$

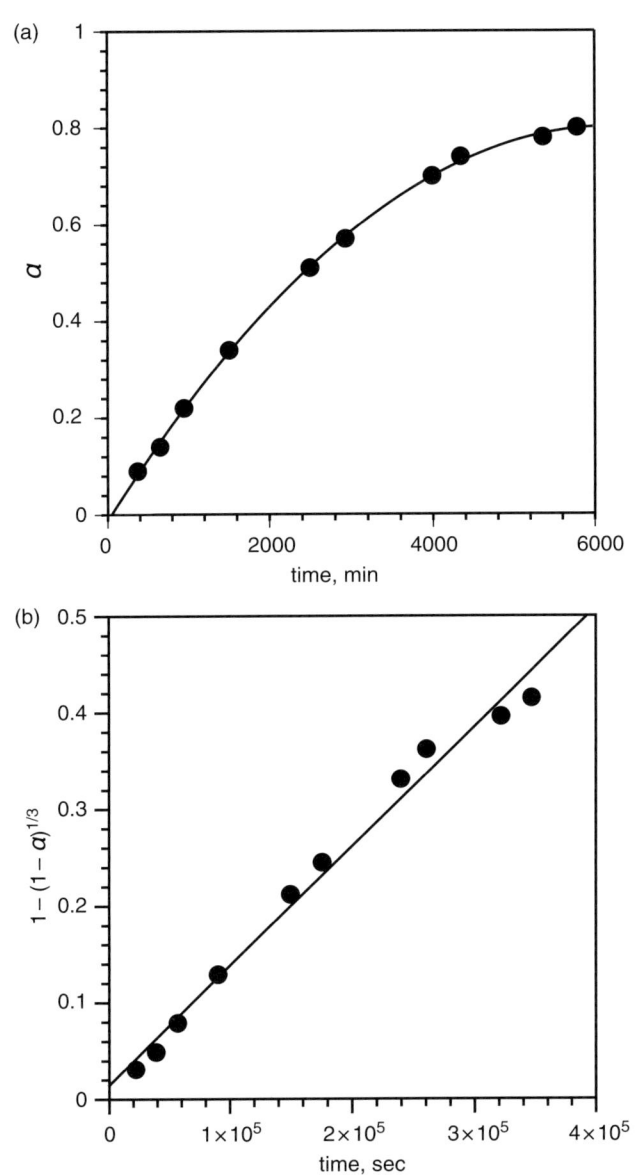

Figure 6.4. (a) Fraction of forsterite reacted away as a function of time. (b) The reacted fraction data fit to the shrinking particle model. The slope of the line is k_p.

So $R_o = 57.5 \times 10^{-6}$ m and $V_m = 4.365 \times 10^{-5}$ m³/mol. Equation (6.52) can be used to find k_+.

$$k_+ = \frac{R_o k_p}{V_m} = \frac{(57.5 \times 10^{-6})(1.23 \times 10^{-6})}{(4.365 \times 10^{-5})} = 1.62 \times 10^{-6} \frac{\text{mol}}{\text{m}^2 \text{ sec}} \qquad (6.54)$$

Particle lifetime model

Geochemists are more familiar with the closely related particle lifetime model of Lasaga (1998b), which is based on the same assumptions as the shrinking particle model. This section shows how these models are related.

When the temperature, solution composition, and other rate-controlling variables remain nearly constant, the dissolution flux ($J = k_+$, mol/m²sec) for a dissolving particle is constant and the release rate of the dissolving substance (r, mol/sec) is proportional to the particle's surface area (A, m²).

$$r = A k_+ \tag{6.55}$$

If the particle is a sphere with a diameter D (m) its area $A = \pi D^2$, m². The rate of volume loss of the particle is the release rate multiplied by the molar volume of the dissolving substance (V_m, m³/mol).

$$\frac{dV}{dt} = -r V_m = -\pi D^2 k_+ V_m \tag{6.56}$$

The rate of volume loss of the particle is also given by the time derivative of the volume change for a sphere ($V = \pi D^3/6$).

$$\frac{dV}{dt} = \frac{\pi D^2}{2} \frac{dD}{dt} \tag{6.57}$$

Setting these two equations equal to each other gives the rate of reduction of the particle's diameter.

$$\frac{dD}{dt} = -2 k_+ V_m \tag{6.58}$$

This equation is integrated and evaluated between the initial diameter (D_o, m) and the diameter (D_t, m) when time = t.

$$D_t = D_o - 2 k_+ V_m t \tag{6.59}$$

If the particle dissolves until its diameter reaches zero, it has dissolved away completely and the particle lifetime (t_ℓ, sec) is found by setting $D_t = 0$ in Eq. (6.59).

$$t_\ell = \frac{D_o}{2 k_+ V_m} \tag{6.60}$$

Because $D_o/2 = R_o$, the lifetime of the grain can also be expressed in terms of k_p as defined by Eq. (6.48). This relationship links the shrinking particle and particle lifetime models.

$$t_\ell = \frac{R_o}{k_+ V_m} = \frac{1}{k_p} \tag{6.61}$$

Equation (6.60) can be rearranged to find the ratio of the original diameter (D_o) to the lifetime (t_l).

$$2k_+ V_m = \frac{D_o}{t_\ell} \tag{6.62}$$

Substituting Eq. (6.62) into Eq. (6.59) gives a relationship between dimensionless diameter and dimensionless time.

$$\frac{D_t}{D_o} = \left(1 - \frac{t}{t_\ell}\right) \tag{6.63}$$

This equation shows that the particle diameter shrinks as a linear function of elapsed time as shown in Figure 6.5.

The surface area of the particle is πD^2, so the diameter can be expressed in terms of the surface area and that relationship can be substituted into Eq. (6.63) to find the reduced surface area as a function of reduced time.

$$D = \left(\frac{A}{\pi}\right)^{1/2} \tag{6.64}$$

$$\frac{A_t}{A_o} = \left(1 - \frac{t}{t_\ell}\right)^2 \tag{6.65}$$

Equation (6.55) states that the rate of transfer of material (r, mol/sec) from the particle to the solution is directly proportional to the surface area, so it can be combined with Eq. (6.65) to give the dimensionless release rate as a function of dimensionless time.

$$\frac{r}{r_o} = \left(1 - \frac{t}{t_\ell}\right)^2 \tag{6.66}$$

The particle diameter can be expressed in terms of the particle volume ($V = \pi D^3/6$) and that relationship can be substituted into Eq. (6.63) to find the reduced particle volume as a function of reduced time.

$$D = \left(\frac{6V}{\pi}\right)^{1/3} \tag{6.67}$$

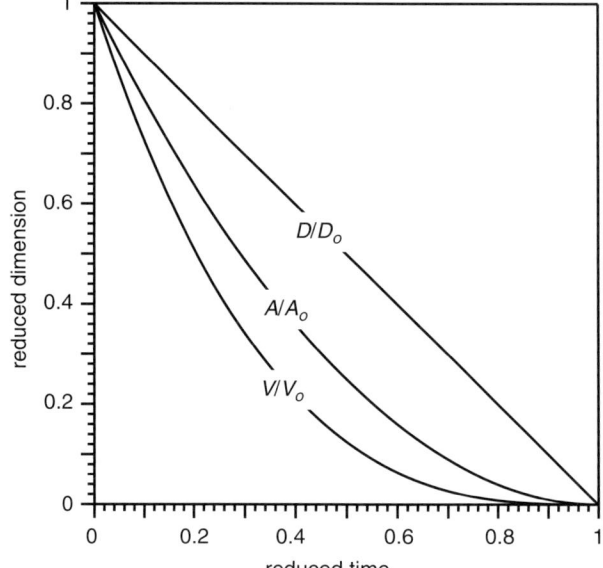

Figure 6.5. Dimensionless diameter, surface area, and volume of a dissolving spherical particle as a function of dimensionless time based on Eqs (6.1), (6.3), and (6.6).

$$\frac{V_t}{V_o} = \left(1 - \frac{t}{t_\ell}\right)^3 \tag{6.68}$$

Multiplying the volume of the particle by the molar volume of the substance gives the number of moles of substance in the particle and the ratio, n_t/n_o, is the fraction of material remaining (p) at any time and p equals one minus the fraction of material that has dissolved away (α).

$$\frac{n_t}{n_o} = p = (1-\alpha) = \left(1 - \frac{t}{t_\ell}\right)^3 \tag{6.69}$$

Combining Eqs (6.61) and (6.69) gives a relationship between α and t.

$$1 - \alpha = (1 - k_p t)^3 \tag{6.70}$$

Taking the cube root of both sides of this equation followed by rearrangement produces the shrinking particle model derived earlier as Eq. (6.51).
Equation (6.63) can also be rearranged and expressed in terms of k_p.

$$\left(1 - \frac{D_t}{D_o}\right) = \frac{t}{t_\ell} = k_p t \tag{6.71}$$

This relationship can be used for dissolving grains that are sampled and their diameters measured over the course of an experiment. The slope of a graph of $(1 - D_t/D_o)$ versus time is k_p ($= 1/t_l = 2k_+V_m/D_o$), which can be used to find the dissolution rate constant. Liu *et al.* (2008) used this approach to determine the rate of nonoxidative dissolution of galena in an acid solution.

Equation (6.66) can be rearranged and expressed in terms of k_p.

$$1 - \left(\frac{r_t}{r_o}\right)^{1/2} = \frac{t}{t_\ell} = k_p t \tag{6.72}$$

If the rate of change of the solution concentration is measured at the onset of the experiment and again at several later times, the slope of a graph of the left-hand side of this equation versus time is k_p ($= 1/t_l = 2rV_m/D_o$), which can be used to find k_p and the dissolution rate constant (k_+). This approach would be especially suitable for mixed flow reactor experiments where the rates are determined directly from the solution composition and flow rate.

Example 6.7. Mineral grain lifetimes

Goldich (1938) proposed a mineral weathering stability series that is based on Bowen's reaction series. According to the Goldich series, olivine is very unstable in the weathering environment and quartz is very stable. Goldich's qualitative model seems reasonable, but we now have extensive mineral dissolution data that can be used to quantify this model by calculating the lifetime of 1 mm diameter mineral grains.

The dissolution rate constant for forsterite at pH = 4.5 is 2.75×10^{-9} mol/m²sec (Palandri and Kharaka, 2004) and molar volume is 4.365×10^{-5} m³/mol (Robie and Hemmingway, 1995). The lifetime of a 1 mm diameter grain is calculated using Eq. (6.60).

$$t_\ell = \frac{D_o}{2k_+V_m} = \frac{1 \times 10^{-3}\,\text{m}}{2\left(2.75 \times 10^{-9}\,\frac{\text{mol}}{\text{m}^2\text{sec}}\right)\left(4.365 \times 10^{-5}\,\frac{\text{m}^3}{\text{mol}}\right)} = 8.33 \times 10^9\,\text{sec} \tag{6.73}$$

The lifetime of a 1 mm diameter forsterite grain is 131 years.

The dissolution rate constant for quartz is 2.75×10^{-13} mol/m²sec (Palandri and Kharaka, 2004) and molar volume is 2.269×10^{-5} m³/mol (Robie and Hemmingway, 1995). The lifetime of a 1 mm diameter grain is calculated using Eq. (6.60).

$$t_\ell = \frac{D_o}{2k_+V_m} = \frac{1 \times 10^{-3}\,\text{m}}{2\left(2.75 \times 10^{-13}\,\frac{\text{mol}}{\text{m}^2\text{sec}}\right)\left(2.269 \times 10^{-5}\,\frac{\text{m}^3}{\text{mol}}\right)} = 8.01 \times 10^{13}\,\text{sec} \tag{6.74}$$

The lifetime of a 1 mm diameter quartz grain is 2.54 million years, which means that the quartz grain will persist 19,000 times longer than the forsterite grain.

Precipitation and dissolution surfaces

Precipitation and dissolution reactions at mineral surfaces are quite complex because both the surface structure and the reaction mechanisms change as a function of the driving force for the reaction. A number of conceptual and quantitative models capture one or another aspect of the growth and/or dissolution process but, like the blind men and elephant proverb, each model explains only selected aspects of these processes so we lack an overall picture of the beast. The point of this section is to caution the reader against over-interpreting any of these models.

Equilibrium condition

At equilibrium the chemical potential of each component in the solid equals its chemical potential in the solution so $\Delta\mu_r = 0$ and the concentration of each component in solution remains unchanged over time. However, dissolution and precipitation reactions continue to occur and their rates are exactly equal because of the principle of detailed balance. This means that although there is no net mass transfer between the solid and solution, these reactions constantly reconstruct the surface. For most solids at room temperature this effect is small but, at the high temperatures or for the long times characteristic of many geological settings, surface reconstruction can transfer significant quantities of trace element or isotopic species between the surface and the solution. At low temperatures the rate of exchange of components between the surface and the interior of the crystal is limited by slow solid-state diffusion, so the composition of crystal interiors typically remains unchanged over geologically significant times. The constant interchange of components between the solution and the surface allows the surface defects and heterogeneities to anneal away over time so that the solid surface evolves toward its equilibrium structure and composition.

The Wulff theorem states that the surface free energy of a crystal face is directly proportional to the distance from the Wulff point, which lies in the interior of the crystal (Mutaftschiev, 2001; Wulff, 1977). For amorphous solids, the equilibrium shape is a sphere. For crystals, the equilibrium shape can be predicted using the Wulff construction, which relates the surface free energy of a crystal face to the length of a vector that is perpendicular to that crystal face and passes through the Wulff point in the interior of the crystal. Wolff and Gualtieri (1962) describe the methods used to predict the equilibrium shape of a crystal. They also point out that crystals seldom achieve a perfect equilibrium shape because they grow from a non-equilibrium, supersaturated solution. Dissolution rate experiments typically use broken grains produced by crushing. The surface free energy of some sites on these grains can be quite high and these sites dissolve quickly when first exposed to the solution (Petrovich, 1981a, 1981b). As dissolution proceeds, these highly reactive surface sites dissolve away and the dissolution rate decreases. However, dissolution continues to deplete the most reactive sites on the surface leading to a surface structure and composition that is unlike the equilibrium surface.

The Kossel crystal model is another useful way to visualize the relationship between crystal structure and crystal shape. This model visualizes a crystal as an ordered assemblage of cube-shaped growth units (Figure 6.6). The Kossel crystal model provides the conceptual foundation for the periodic bond chain theory (Hartman and Perdok, 1955a, 1955b, 1955c). Periodic bond chains (PBC) are chains of strong bonds that run through the crystal and define the direction of a PBC vector. Flat faces (F-faces) on a Kossel crystal are parallel to at least two PBC vectors. Stepped faces (S-faces) are parallel to at least one PBC vector and kinked faces (K-faces) are not parallel to any PBC vector.

Whether at equilibrium or not, the surface atoms of a solid are underbonded so they must complete their coordination by forming bonds with H_2O, OH, or a ligand from the solution. The surface species can ionize in response to changing pH to produce a surface charge that is responsible for further interactions with solution species. Stumm (1992) summarizes the most important models of the chemistry of the solid/solution interface.

Slightly undersaturated/slightly supersaturated conditions

Departure from equilibrium produces a growing or dissolving surface with a composition and structure that reflects competing kinetic processes. Growing crystal surfaces develop convex surface features (hillocks), which because of surface free energy effects have a chemical potential that is greater than atomically flat, equilibrium crystal faces. Dissolving crystals develop concave surface structures such as etch pits that have a lower chemical potential than atomically flat, equilibrium faces. The shape of growing crystals is controlled by a complex interplay of growth rate and growth site inhibition by adsorbed species (Sunagawa, 2005). Although less well studied, crystal dissolution shapes appear to be controlled by similar processes. Multicomponent minerals dissolve incongruently as weakly bonded components are preferentially released to the solution. This enriches the surface in strongly bonded components, which makes the dissolving surface behave like a metastable phase that is less soluble than the bulk phase.

The dissolution or precipitation rate for near-equilibrium conditions can be modeled as a function of reaction affinity (A, kJ/mol = $-\Delta\mu_r$).

$$J = J_+\left(1 - e^{-\Delta\mu_r/RT}\right) = J_+\left(1 - e^{A/RT}\right) \tag{6.75}$$

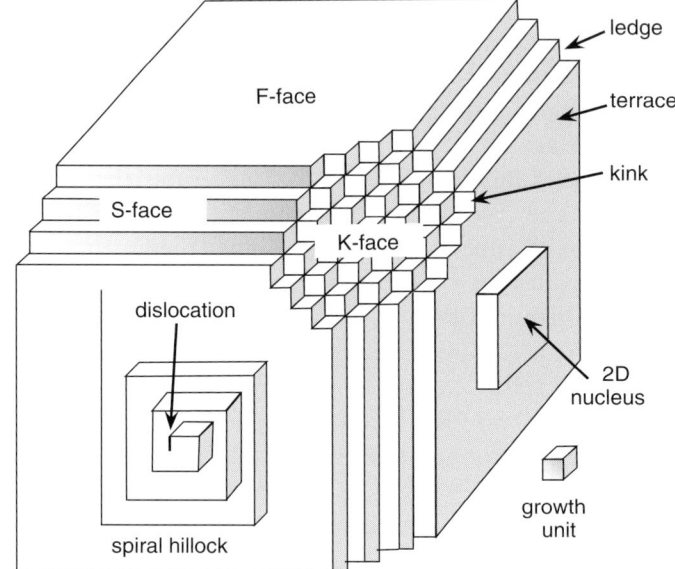

Figure 6.6. Schematic diagram of a Kossel crystal showing F-, S-, and K-faces predicted by the PBC theory along with features predicted by various growth models. Based on Figure 1 in Sleutel *et al.* (2012).

The affinity model is based on the principle of detailed balance (see the derivation in Chapter 3). According to this model, when $A = 0$, $J = 0$ and when $A > 5RT$, $J \approx J_+$. This model seems to work for simple compounds, such as quartz, that dissolve congruently and reversibly. It is easier to visualize the goodness of fit by converting Eq. (6.75) to a linear form (Figure 6.7).

$$(J/J_+) = 1 - e^{A/RT} \qquad (6.76)$$

Although this model has considerable conceptual value and provides modelers with a simple way to force reaction rates to zero at equilibrium, it is not well suited to express the complexities associated with the dissolution of many minerals. Changes in the reaction mechanism with increasing affinity (Arvidson and Luttgbe, 2010) reset the far-from-equilibrium dissolution flux, J_+, to different values so the J versus A graph does not approach a plateau at $A = 5RT$. Incongruent dissolution changes the surface composition away from that of the bulk solid (Schott and Oelkers, 1995) so the chemical potential of the surface components are different from the bulk solid. This means that the $A = 0$ reference point based on the unreacted mineral does not apply to the leached surface and the J versus A graph does not cross the affinity axis at zero. Both of these effects invalidate the fundamental assumption of the principle of detailed balance, which requires that the forward and reverse reactions remain the same over the entire range of the model.

Slightly undersaturated/slightly supersaturated conditions

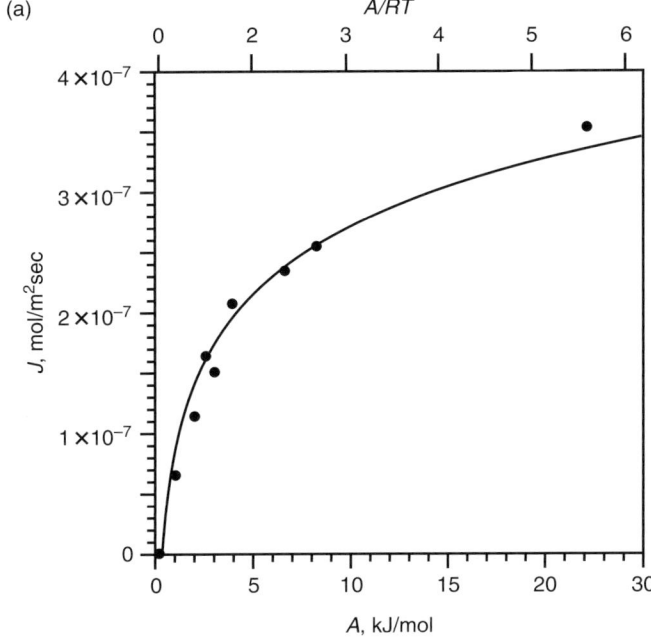

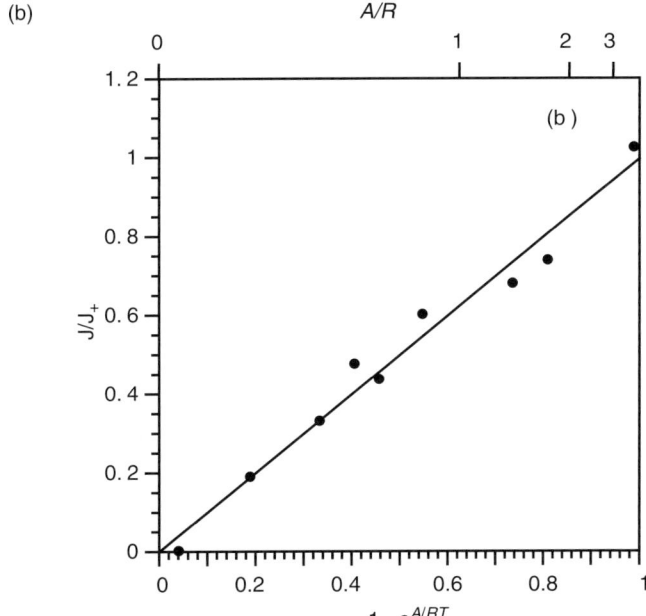

Figure 6.7. (a) Quartz dissolution flux at 300°C, J, as a function of the affinity, A, driving the reaction. (b) Quartz dissolution flux, J/J_{+}, as a function of the linearized affinity, $1 - e^{A/RT}$. The slope of the best-fit solid line is 0.88 and the theoretical slope is 1.0. Data from Berger et al. (1994).

The Kossel crystal model shown in Figure 6.6 provides the conceptual basis for many models of crystal growth and dissolution. Most of those models also incorporate the concepts of terrace, ledge, and kink (TLK), which describe the free energy of attachment of a growth unit as being proportional to the number of bonds (1 through 6) between it and the crystal. According to the TLK model, the energy needed to form a 2D nucleus, which initiates the growth of a new layer, is high, making the predicted rates of crystal growth much lower than is observed. This problem was resolved by the Burton–Cabrera–Frank (BCF) theory (Burton and Cabrera, 1949; Burton et al., 1949, 1951; Cabrera and Burton, 1949) that postulates that new ledges are the result of line dislocations that grow into spiral hillocks (Figure 6.6). Zhang and Nancollas (1990) and Teng et al. (2000) give examples of the application of these concepts to the growth of minerals. Dissolution processes can be visualized as the inverse of the Kossel crystal growth processes. For example, line defects at the spiral hillocks provide steps that enhance dissolution rates leading to etch pit formation (Brantley et al., 1986a; Brantley et al., 1986b). Etch pits and vacancy islands (Dove et al., 2005) exert a profound control on the surface structure of dissolving solids (Lasaga and Blum, 1986; MacInnis and Brantley, 1993). Etch pits and the opening of vacancy islands cause weathered mineral surfaces to be extremely rough (Berner et al., 1980; Velbel, 1989). Monte Carlo models of dissolving Kossel crystal surfaces show that surface roughness increases with the extent of the reaction and the departure from equilibrium (Bandstra and Brantley, 2008; Wehrli, 1989). Because ledge retreat can initiate at crystal edges, the edge regions dissolve more quickly than the centers of the faces, so the dissolving crystal becomes rounded. Eventually the dissolving crystal becomes bounded by rough and curved surfaces from which growth units can be removed at any location (Cabrera and Vermiyea, 1958).

Incongruent dissolution changes the surface chemistry of dissolving minerals. Incongruent dissolution occurs when there are significant differences among the bond strengths of the mineral's components. This means that in the initial stages of the dissolution process, weakly bonded components are released quickly and strongly bonded components are retained in the surface. The release rate of the weakly bonded components diminishes as their concentration declines until their release rate matches the release rate of the strongly bonded components. Eventually the dissolution becomes congruent as the mineral surface becomes paved by a "leached layer" that is enriched in the slowest dissolving components (Hellmann et al., 1990; Schott et al., 2012). The leached layer can undergo reconstruction reactions that cross-link the components, which further slows its dissolution rate (Casey et al., 1993).

Very undersaturated/very supersaturated conditions

With increasing supersaturation, dissolved species interact to form oligomers that increase in size in proportion to the degree of supersaturation. These oligomers become the most important growth units for precipitating solids (Cölfen and Antonietti, 2008). The oligomers undergo self-assembly and their shape has a profound effect on the internal structure of the growing solid (Damasceno *et al.*, 2012). Oligomer attachment causes microscopic surface roughening and their slow diffusion rates promote the formation of dendrites. Dendritic ice crystals, e.g. snowflakes, are a familiar example of this growth habit (Furukawa and Shimada, 1993). There are many examples of mineral dendrite formation.

Incongruent dissolution and leached layer thickness becomes more manifest at very low pH (Weissbart and Rimstidt, 2000). Under these conditions the leached layer can dissolve by shedding oligomers as well as monomers (Dietzel, 2000; Weissbart and Rimstidt, 2000). These oligomers subsequently depolymerize after they move away from the dissolving surface.

Chapter 7
Diffusion and advection

Diffusion and fluid advection are the most important mass transfer processes in low temperature geochemical systems. Whenever mass transfer processes are slow relative to chemical reaction rates, diffusion and advection processes will control the overall rate of a geochemical process. This chapter introduces some simple models that link transport rates to chemical reaction rates.

Advection

Advection rates are expressed in terms of volume discharge, specific discharge, or surficial velocity depending upon the requirements of the model. Discharge (also called volumetric flow rate) ($Q = dV/dt$, m^3/sec) is the volume of fluid passing through a cross-sectional area per unit time. Specific discharge (also called Darcy velocity) ($q = Q/A$, m/sec) is the volumetric flow rate divided by the surface area through which the fluid is flowing. Surficial velocity (v, m/sec) is the fluid velocity relative to an adjacent solid surface. For a packed bed, the surficial velocity is the specific discharge divided by the porosity ($v = q/\varphi$).

Viscosity is a fluid's resistance to flow, so advection models must account for viscosity either as an explicit term or in a scaling constant. The dynamic viscosity (η, kg/m sec) equals the kinematic viscosity (ν, m^2/sec) divided by the density (ρ, kg/m^3). The dynamic viscosity of liquid water in equilibrium with the vapor phase for temperatures between the freezing and critical points can be computed from an equation given by Watson *et al.* (1980).

$$\ln \eta = -10.1357 + 1.5064\tau^{1/3} - 0.1189\tau + 1.7604\tau^2 - 0.53367\tau^{7/3} + 4.9643 \times 10^{-3}\tau^{10}$$
$$\tau = \frac{647.073}{T(\text{K})} - 1 \tag{7.1}$$

The density of liquid water in equilibrium with the vapor phase for temperatures between the freezing and critical points can be computed from an equation given by Wagner and Pruß (2002).

$$\rho = 322\begin{pmatrix} 1+1.9927\vartheta^{1/3} +1.0997\vartheta^{2/3} - 0.51089\vartheta^{5/3} \\ -1.7549\vartheta^{16/3} - 45.517\vartheta^{43/3} - 6.7469\times 10^5 \vartheta^{110/3} \end{pmatrix} \qquad (7.2)$$

$$\vartheta = 1 - \frac{T(K)}{647.096}$$

Fluids flow in response to a pressure difference. Buoyancy forces, due to density differences related to differences in the temperature or salinity, can cause fluid flow. Buoyancy forces are considered in models of convective flow. Fluids also flow because of differences in the hydrostatic head between a source and discharge region. Hydrostatic head is the difference in elevation (Δz, m), which produces a pressure difference because of gravitational acceleration ($P = \rho g \Delta z$). The Manning equation and Darcy's law are examples of equations that model flow driven by hydrostatic forces. The Manning equation predicts flow velocity (v, m/sec) in open channels (Chaudhry, 2008) as a function of the channel's cross-sectional area (A, m^2), wetted perimeter (P, m), and the slope of the water surface (s, m/m).

$$v = \frac{k}{n}(A/P)^{2/3} s^{1/2} \qquad (7.3)$$

In this model, k is a flow constant (= 1 m$^{1/3}$/sec for SI units) that makes the units work out and n (no units) is an empirically determined channel roughness factor that ranges from ~0.03 for clean and straight natural streams to ~0.05 for natural streams lined with cobbles and large boulders (Barnes, 1967; Limerinos, 1970). More sophisticated models have been developed and calibrated (Bjerklie and Dingman, 2005). Darcy's law predicts the specific discharge (q, m/sec) in porous media (packed beds) as a function of the pressure drop across the bed (ΔP, Pa) and the length of the bed (L, m).

$$q = \left(\frac{k}{\eta}\right)\left(\frac{\Delta P}{L}\right) \qquad (7.4)$$

In this model, k (m^2) is the permeability of the packed bed and η (kg/m sec) is the dynamic viscosity of the fluid.

Dimensionless numbers

Dimensionless numbers are commonly used in diffusion and advection models because they simplify the scaling of the models from laboratory experiments to practical dimensions. This approach also takes advantage of the Buckingham π theorem (Barenblatt, 2003), which states that n independent variables with k independent dimensions can be expressed in terms of p independent dimensionless numbers.

$$p = n - k \tag{7.5}$$

Dimensionless numbers are widely used in engineering disciplines but are less common in geochemical models. The notation for dimensionless numbers consists of a capitalized first letter followed by a lower case letter. Methods used to estimate the numerical values of dimensionless numbers range from purely theoretical to purely empirical. It is advisable to carefully review each estimation method before incorporating a dimensionless number into a model.

Dimensionless numbers are scale independent and contain a characteristic length term (L) that adjusts their magnitude to match the scale of the model. The velocity, v (m/sec), in these definitions is the superficial velocity of the fluid.

The Reynolds number, Re, is defined as the ratio of inertial forces, which resist a change in the flow velocity, to the viscosity, which slows the flow velocity (Denny, 1993; Vogel, 1994). Purcell (1977) discusses several applications of the Reynolds number to real situations. For flow past a flat plate the Reynolds number is

$$\text{Re} = \frac{\rho v L}{\eta} \tag{7.6}$$

For this geometry, when Re < ~2300 the flow is laminar and when Re > ~4000 the flow is turbulent.

For flow through a packed bed with a porosity (ϕ, no units) and consisting of grains with a diameter (D, m), the Reynolds number is

$$\text{Re} = \frac{\rho v D}{\eta(1-\phi)} \tag{7.7}$$

When Re < ~10 the flow is laminar and when Re > ~2000 the flow is fully turbulent.

For flow in a smooth-walled fracture (Qian *et al.*, 2005) with an aperture width of L, the Reynolds number is

$$\text{Re} = \frac{\rho v L}{2\eta} \tag{7.8}$$

When Re > ~2300 the flow is turbulent.

For a vessel stirred (Lamberto *et al.*, 1999) with an agitator of diameter (D, m) with a rotational speed of N (revolutions/sec), the Reynolds number is

$$\text{Re} = \frac{\rho N D^2}{\eta} \tag{7.9}$$

When Re > ~10,000 the flow is fully turbulent.

The Schmidt number, Sc, is defined as the ratio of viscosity to diffusivity (D_i, m²/sec).

$$Sc = \frac{\eta}{\rho D_i} = \frac{\nu}{D_i} \quad (7.10)$$

The Schmidt number and Reynolds number can be combined to find the Péclet number.

$$Pe = Re\,Sc \quad (7.11)$$

The Péclet number is defined as the ratio of mass transfer by advection (v, m/sec) to mass transfer by dispersion (D_L, m²/sec) and diffusion (D_i, m²/sec).

$$Pe = \frac{vL}{D_i + D_L} \quad (7.12)$$

For Pe < 50, diffusion/dispersion dominates the transfer of material between the solution and surface; and for Pe > 100, advection dominates. For many geochemical scenarios, $D_i \ll D_L$ (Knapp, 1989) and only longitudinal dispersion needs to be considered in models. For Taylor dispersion in a fracture with a half-width of L (m), the longitudinal dispersion coefficient is (Horne and Rodriguez, 1983)

$$D_L = \frac{2}{105} \frac{L^2 v^2}{D_i} \quad (7.13)$$

The longitudinal dispersivity, α_L (m), is the longitudinal dispersion coefficient (D_L, m²/sec) plus the diffusion coefficient (D_i, m²/sec) divided by the fluid velocity (v, m/sec).

$$\alpha_L = \frac{D_L + D_i}{v} \quad (7.14)$$

This makes the Péclet number inversely related to the longitudinal dispersivity (Knapp, 1989).

$$Pe = \frac{1}{\alpha_L} \quad (7.15)$$

The 1 in the numerator of this fraction has units of meters. Empirical studies of flow in geological media have demonstrated that α_L is related to the scale of the study, L (m) (Table 7.1).

Table 7.1. *Values of* c *and* m *(Schulze-Makuch, 2005) that can be used to calculate longitudinal dispersivity using Eq. (7.16).*

Medium	c	m
Unconsolidated	0.085	0.81
Sandstones	0.01	0.92
Carbonates	0.80	0.40
Basalts	0.15	0.61
Granites	0.21	0.51

$$\alpha_L = cL^m \tag{7.16}$$

The Damköhler numbers (Da) compare the rate of consumption or production of a dissolved substance to the rate of delivery or removal of that substance at the reaction site (Steefel, 2008). The first and second Damköhler numbers are most useful for geochemical models. Da_I compares the reaction rate to the advection rate.

$$Da_I = \frac{(A/M)kL}{m_{eq}v} \tag{7.17}$$

If $Da_I > 1$, the reaction produces or consumes aqueous species faster than they are delivered or removed by advection, so the overall rate is limited by fluid flow. Da_{II} compares the reaction rate to the rate of diffusive mass transfer across the solid/solution interface.

$$Da_{II} = Da_I Pe = \frac{(A/M)kL^2}{D_i m_{eq}} \tag{7.18}$$

If $Da_{II} > 1$, the reaction produces or consumes aqueous species faster than they are delivered or removed by diffusion so the reaction rate is limited by diffusion. A/M (m²/kg) is the ratio of the interfacial area between the solution and solid to the mass of solution; k (mol/m²sec) is the dissolution flux rate constant; m_{eq} (mol/kg) is the equilibrium concentration; v (m/sec) is the linear flow velocity; and D_i (m²/sec) is the effective diffusion coefficient.

Fick's laws

Diffusion is caused by the displacement of molecules due to random molecular motions. If the molecules are not homogeneously distributed in a region,

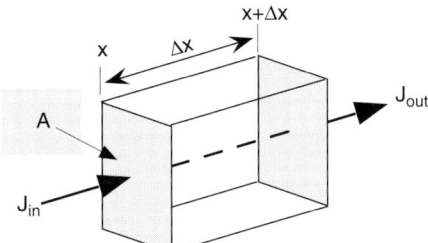

Figure 7.1. Volume element used to develop Fick's second law.

more of them will be displaced away from areas of high concentration into areas of low concentration than vice versa. This causes net transport down a concentration gradient. Erdey-Grúz (1974) and Robinson and Stokes (1959) provide theoretical details about how and why diffusion occurs in aqueous solutions. Cussler (2009), Denny (1993), Levich (1962), Probstein (1989), and Vogel (1994) are useful resources for models that link diffusion and advection. Berner (1980) and Lerman (1979) present models that are applicable to sediments.

Fick's laws are the basis for diffusive transport models. Fick's first law states that diffusive flux (J, mol/m²sec), is the product of a concentration gradient (dc/dx, mol/m⁴) and a diffusion coefficient (D, m²/sec). Note that concentration (c) is expressed with units of mol/m³.

$$J = -D\frac{dc}{dx} \tag{7.19}$$

Fick's second law is related to the first by a simple derivation, which considers the volume element shown in Figure 7.1. As molecules diffuse into and out of the volume element, the rate of change in the number of molecules in the volume is the difference between the flux into and out of the volume across the A areas at each end of the volume.

$$\frac{\Delta n}{\Delta t} = -A\left(J_{x+\Delta x} - J_x\right) \tag{7.20}$$

The change in the number of molecules divided by the volume of the element is the concentration change in the volume.

$$\Delta c = \frac{\Delta n}{A\Delta x} \text{ so } \Delta n = \Delta c\left(A\Delta x\right) \tag{7.21}$$

Substituting Eq. (7.21) into Eq. (7.20) gives the rate of change of concentration in the volume.

Diffusion and advection

$$\frac{\Delta c(A\Delta x)}{\Delta t} = -A(J_{x+\Delta x} - J_x) \tag{7.22}$$

$$\frac{\Delta c}{\Delta t} = -\left(\frac{A}{A}\right)\frac{(J_{x+\Delta x} - J_x)}{\Delta x} = -\frac{(J_{x+\Delta x} - J_x)}{\Delta x} \tag{7.23}$$

The fluxes are defined in terms of the concentration gradients using Fick's first law.

$$\frac{\Delta c}{\Delta t} = -\frac{\left(D\left(\frac{\partial c}{\partial x}\right)_{x+\Delta x} - D\left(\frac{\partial c}{\partial x}\right)_x\right)}{\Delta x} = -\frac{D\left(\left(\frac{\partial c}{\partial x}\right)_{x+\Delta x} - \left(\frac{\partial c}{\partial x}\right)_x\right)}{\Delta x} \tag{7.24}$$

When $\Delta x \to 0$, Eq. (7.24) becomes Fick's second law.

$$\frac{dc}{dt} = -D\left(\frac{d^2c}{dt^2}\right) \tag{7.25}$$

Using Fick's laws to model geochemical situations requires the modeler to define boundary conditions based on the geometric and chemical relationships between the solids and the solution. The resulting differential equations can be quite challenging to solve but many analytical solutions have already been developed. Crank (1975) gives detailed solutions for many relevant problems and because Fourier's law is the analog of Fick's first law, with suitable replacement of variables, the solutions for heat flow problems in Carslaw and Jaeger (1959) can be applied to diffusion models.

Diffusion coefficients

Several experimental designs are used to measure diffusion coefficients (Robinson and Stokes, 1959). Tracer diffusion coefficients for ions are typically calculated from the limiting equivalent conductivity of the ions (Lerman, 1979; Oelkers and Helgeson, 1988; Robinson and Stokes, 1959). At 25°C, the diffusion coefficients for most ions range between 0.5×10^{-9} and 10×10^{-9} m^2/sec (Li and Gregory, 1974; Miller, 1982; Oelkers and Helgeson, 1988). The diffusion coefficients for uncharged organic species are similar to those for most ions (Oelkers, 1991).

The diffusion coefficients of H^+ and OH^- ions are much larger than other ions because their movement in solution involves a kind of hopping mechanism (Grotthuss mechanism) related to the very fast dissociation and re-association of water molecules (Cukierman, 2006; Tuckerman et al., 2002).

Table 7.2. *Diffusion coefficients for selected aqueous species (Miller, 1982; Oelkers and Helgeson, 1988). The* a, b, *and* c *values are coefficients for Eq. (7.27).*

Species	D (25°C) m²/sec	a	b	c	E_a (25°C) kJ/mol	E_a (125°C) kJ/mol
H^+	9.3×10^{-9}	−7.330	194.7	−119613	11.6	4.3
OH^-	5.5×10^{-9}	−7.418	278.3	−158688	15.1	5.3
Na^+	1.3×10^{-9}	−6.990	−196.5	−107366	17.5	10.4
Ca^{2+}	0.9×10^{-9}	−7.585	177.9	−185795	20.5	9.6
Cl^-	2.0×10^{-9}	−7.049	−96.55	−115838	16.7	9.0
SO_4^{2-}	1.1×10^{-9}	−6.959	−256.2	−100951	17.9	11.7
methane	1.7×10^{-9}	−6.849	−203.5	−109402	18.0	11.2
octane	0.7×10^{-9}	−7.212	−223.7	−104675	17.7	11.3
methanol	1.6×10^{-9}	−6.984	−70.83	−140820	19.5	10.8
octanol	0.7×10^{-9}	−7.335	−41.29	−147114	19.7	10.6
benzene	1.1×10^{-9}	−7.262	63.43	−171300	20.8	10.2
acetic acid	1.2×10^{-9}	−7.026	−150.8	−122463	18.6	11.1

Effect of temperature, solution composition, and tortuosity on *D*

Diffusion is a thermally activated process so diffusion rates increase with increasing temperature. For many types of diffusion, this temperature effect is modeled by fitting the diffusion coefficient to an Arrhenius type law.

$$D = Ae^{-E_a/RT} \tag{7.26}$$

For aqueous species, graphs of log *D* versus $1/T$ are curved because E_a decreases with increasing temperature (Figure 7.2). Values of E_a range from around 20 kJ/mol at 25°C to near 10 kJ/mol at 300°C (Table 7.2). E_a for H^+ and OH^- are lower and their diffusion coefficients change somewhat less with increasing temperature, but they are always larger than for other species. Diffusion coefficients for aqueous species along the water liquid/vapor curve are usually fit by a polynomial in $1/T$.

$$\log D = a + \frac{b}{T} + \frac{c}{T^2} \tag{7.27}$$

This makes the activation energy a function of temperature.

$$E_a = 2.303R\left(\frac{d \log D}{d(1/T)}\right) = 2.303R\left(b + \frac{2c}{T}\right) \tag{7.28}$$

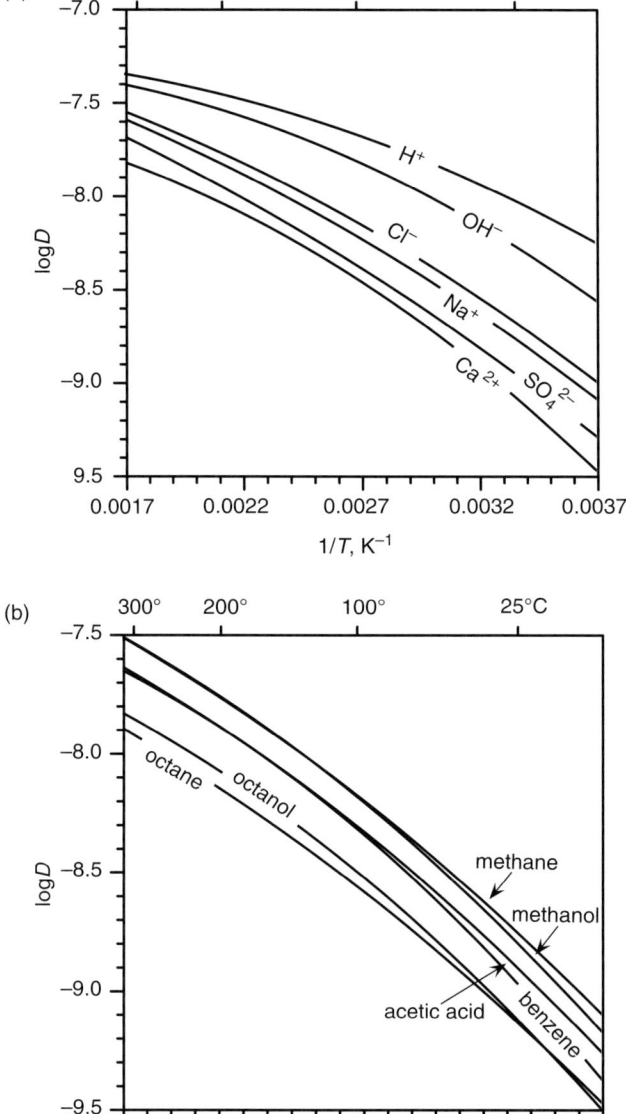

Figure 7.2. Log D versus $1/T$ behavior for selected aqueous species (Miller, 1982; Oelkers, 1991; Oelkers and Helgeson, 1988).

According to the Stokes–Einstein equation, diffusion coefficients have a reciprocal relationship to the dynamic viscosity of the host fluid (η) and the hydrodynamic radius of the diffusing species (R).

$$D = \frac{K_B T}{6\pi \eta R} \quad (7.29)$$

This equation can be recast to create a function that can be used to extrapolate diffusion coefficients to other temperatures.

$$D_{T_2} = \left(\frac{\eta_{T_1}}{T_1}\right)\left(\frac{T_2}{\eta_{T_2}}\right) D_{T_1} \qquad (7.30)$$

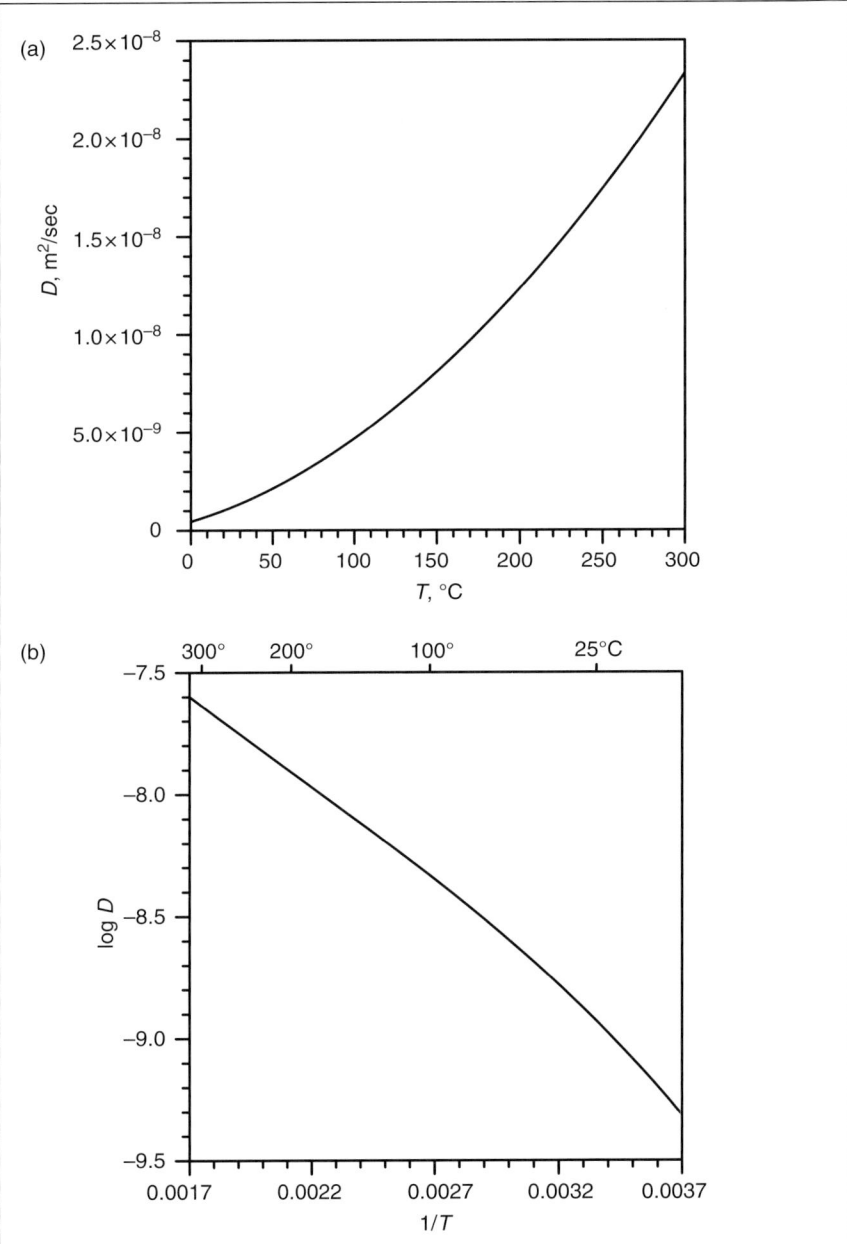

Figure 7.3. (a) Diffusion coefficient of H_4SiO_4 as a function of temperature estimated using the Stokes–Einstein equation. (b) Log D for H_4SiO_4 versus $1/T$ estimated using the Stokes–Einstein equation.

> **Example 7.1.** Diffusion coefficient of silica (H_4SiO_4) as a function of temperature
>
> The diffusion coefficient of H_4SiO_4 in pure water at 25°C is 1.17×10^{-9} m²/sec (Rebreanu et al., 2008). This value can be extrapolated to hydrothermal temperatures using Eq. (7.30). The easiest way to perform this extrapolation is to log transform Eq. (7.30).
>
> $$\log D_{T_2} = \log\left(\frac{\eta_{T_1}}{T_1}\right) + \log\left(\frac{T_2}{\eta_{T_2}}\right) + \log D_{T_1} \quad (7.31)$$
>
> Values of η are calculated from Eq. (7.1) and used to find the first and second terms of Eq. (7.31). The values of $\log(T_2/\eta_{T_2})$ can be fitted to a temperature function.
>
> $$\log\left(\frac{T_2}{\eta_{T_2}}\right) = -100039\left(\frac{1}{T_2}\right)^2 - 289.64\left(\frac{1}{T_2}\right) + 7.6142 \quad (7.32)$$
>
> At $T_1 = 298$°C, $\log(\eta/T_1) = -5.524$ and $\log D_{T_1} = -8.93$. These values along with Eq. (7.32) are inserted into Eq. (7.31) to produce a temperature function for $\log D_{T_2}$.
>
> $$\log D_{T_2} = -100039\left(\frac{1}{T_2}\right)^2 - 289.64\left(\frac{1}{T_2}\right) - 6.840 \quad (7.33)$$
>
> The values of log D and D predicted by Eq. (7.33) are shown in a graph in Figure 7.3.

The effect of the solution composition on D is complex. To maintain charge neutrality, the diffusion rates of cations and anions must be linked. The mutual diffusion coefficient for combinations of cations and anions is approximately the harmonic mean of their tracer diffusion coefficients.

$$D_{mutual} = \frac{2 D_{cation} D_{anion}}{D_{cation} + D_{anion}} \quad (7.34)$$

In addition, ions can interact with each other to form ion pairs and complexes. The diffusion coefficients of ion pairs can be estimated using a method described by Applin and Lasaga (1984).

Diffusion rates in porous media are slower than in bulk solution because the species must diffuse around the solid particles, which increases their diffusion path length. This increase in diffusion path length is expressed as tortuosity (τ), which is defined as the ratio of the actual distance traveled by the species to the linear distance between the initial and final location of the species. Tortuosity increases with decreasing porosity (ϕ) and there are many theoretical and empirical models for this relationship (Lerman, 1979; Shen and Chen, 2007). Three examples of empirical fits taken from Shen and Chen (2007) illustrate the diversity of tortuosity models.

$$\tau^2 = \phi^{-1.14} \tag{7.35}$$

$$\tau^2 = \phi + 3.79(1-\phi) \tag{7.36}$$

$$\tau^2 = 1 - 2.02\ln\phi \tag{7.37}$$

Diffusion coefficients can be adjusted for use in porous media by dividing them by τ^2.

$$D_\tau = \frac{D}{\tau^2} \tag{7.38}$$

Mineral dissolution without flow

Fick's second law can be used to model the one-dimensional concentration gradient in a static solution in contact with a dissolving mineral. This model solves Eq. (7.25) for boundary conditions specifying that when $x = 0$, $c = c_s$ and when $x = \infty$, $c = 0$. This model predicts concentration as a function of position and time (Crank, 1975).

$$c(x,t) = c_s \operatorname{erfc}\left(\frac{x}{2\sqrt{Dt}}\right) \tag{7.39}$$

In this equation, erfc(z) is the complementary error function: erfc(z) = 1 − erf(z).

> **Example 7.2.** Gypsum (CaSO$_4$•2H$_2$O) dissolution in an unstirred solution
>
> If gypsum is immersed in an unstirred aqueous solution it will dissolve more rapidly than the Ca^{2+} and SO$_4^{2-}$ ions can diffuse away from the surface. This means that the solution very near the surface will be nearly in equilibrium with the gypsum while the solution distant from the surface will contain no Ca^{2+} or SO$_4^{2-}$ ions. Equation (7.39) can be used to predict the concentration of these ions as a function of distance from the surface at a chosen time.
> The mutual diffusion coefficient of CaSO$_4$ is found using Eq. (7.34).

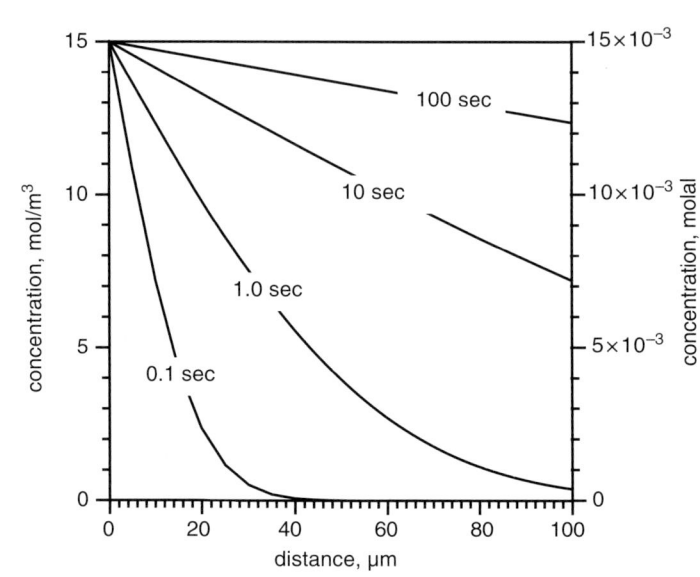

Figure 7.4. Concentration of Ca^{2+} or SO_4^{2-} as a function of distance and elapsed time from a gypsum surface. The gypsum is dissolving into a static solution.

$$D_{CaSO_4} = \frac{2D_{Ca^{2+}}D_{SO_4^{2-}}}{D_{Ca^{2+}} + D_{SO_4^{2-}}} = \frac{2(0.9 \times 10^{-9})(1.1 \times 10^{-9})}{0.9 \times 10^{-9} + 1.1 \times 10^{-9}} = 0.95 \times 10^{-9} \quad (7.40)$$

D_{CaSO_4} is 0.95×10^{-9} m²/sec and gypsum solubility is 15×10^{-3} mol/kg ($c_s = 15$ mol/m³) (Dutrizac, 2002). Using these values, Eq. (7.39) predicts the concentration versus distance profiles shown in Figure 7.4.

Transport-limited dissolution

Reactions are limited by the slowest step in the overall process. If the reaction flux (J_R, mol/m²sec) for a dissolving solid is fast compared to the diffusion flux (J_D, mol/m²sec), the dissolution process is limited by the rate of transfer of reactants or products between the solution and the solid's surface. When $J_R \approx J_D$, the overall rate of dissolution is controlled by mixed kinetics, which is modeled by convolving the models for both the dissolution and transport process.

The second Damköhler number, Eq. (7.18), is often used to test whether a process is limited by diffusive transport. Da_{II} is the ratio of the rate of reaction at the surface (numerator) to the rate of transfer between the surface and the solution (denominator). If $Da_{II} \gg 1$, the surface reaction is fast relative to the diffusion rate. If $Da_{II} \ll 1$, the surface reaction is slow relative to the transport rate, so the overall rate is reaction limited. If Da_{II}

≈ 1, the surface reaction rate is nearly the same as the transport rate, so the overall rate is controlled by mixed kinetics. The Da_{II} test applies only for no flow situations.

Example 7.3. Gypsum dissolution into a static fluid

At Earth surface conditions, gypsum ($CaSO_4 \cdot 2H_2O$) dissolution appears to be controlled partly by the rate of Ca^{2+} and SO_4^{2-} release at the mineral surface and partly by the rate of transport of these species away from the dissolving surface. The second Damköhler number, Eq. (7.18), can be used to determine the relative importance of reaction rate versus diffusion rate for gypsum dissolving into a static solution in a fracture with a width, L, of 1.0×10^{-3} m.

$A/V = 2/L = 2.0 \times 10^3$ m^2/m^3 (Table 6.1)
$k_+ = 7.0 \times 10^{-5}$ mol/m^{2}sec (Colombani, 2008)
$L = 1.0 \times 10^{-3}$ m
$D_{CaSO_4} = 9.54 \times 10^{-9}$ m^2/sec (Example 7.2)
$c_{eq} = 15$ mol/m^3 (Dutrizac, 2002)

$$Da_{II} = \frac{(A/V)k_+L^2}{D_i c_{eq}} = \frac{(2.0 \times 10^3)(7.0 \times 10^{-5})(1.0 \times 10^{-3})^2}{(9.45 \times 10^{-9})(15)} = 0.99$$

This means that for a 1 mm (0.001 m) wide fracture, the rate of gypsum dissolution nearly equals the rate of diffusive transport away from the dissolving surface. For narrower fractures, $Da_{II} < 1$, which means that the rate of diffusive transfer is fast compared to the dissolution rate. $Da_{II} > 1$ for wider fractures, which means that the reaction rate is faster than the diffusive transfer rate.

Mass transfer coefficients are the basis for models where the dissolved species are transported by a combination of diffusive and advective processes. The diffusive mass transfer coefficient (k_D, m/sec) is based on boundary layer theory. The basic premise of boundary layer theory is that, for laminar flow, the fluid velocity adjacent to a solid surface is zero (the "no slip condition") and the velocity increases as a parabolic function of distance away from the surface until it matches the velocity of the bulk fluid (Figure 7.5). This means that there is a thin layer of fluid with a thickness of δ_D (m) adjacent to the surface that is effectively static. The rate of mass transport through this layer is limited by the diffusion rate of the dissolved species. The diffusional boundary layer is much thinner than the velocity boundary layer. For laminar flow past a flat surface, the thickness of the diffusional boundary layer is related to the thickness of the velocity boundary layer (δ_V) by the Schmidt number, which compares the fluid viscosity to the diffusivity (Probstein, 1989).

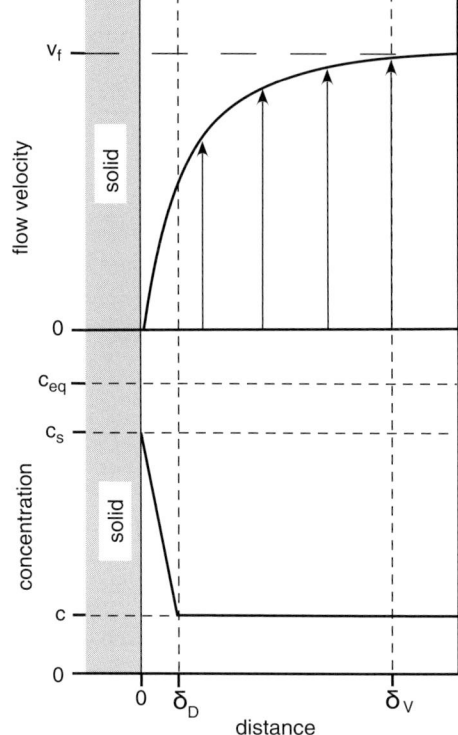

Figure 7.5. The velocity boundary layer thickness (δ_V) and the diffusional boundary layer thickness (δ_D) for laminar flow past a flat surface. The diffusional boundary layer is approximated as a thin layer of static fluid at the solid surface where only diffusional mass transport occurs. See discussions in Probstein (1989), Denny (1993), and Vogel (1994) for more details.

$$\delta_D = \frac{0.6}{Sc^{1/3}} \delta_V \qquad (7.41)$$

The Schmidt number appears in most formulations of diffusion mass transfer coefficients (Table 7.3).

The Nernst film model is used to quantify diffusional transport through the "static" boundary layer. This model approximates the low velocity boundary layer as a thin, static film between the surface and the free flowing solution. Fick's first law gives the diffusional flux (J_D, mol/m²sec) through this film (Figure 7.6).

$$J_D = \frac{D_i}{\delta_D}(c_s - c) \qquad (7.42)$$

The concentration gradient is the surface concentration (c_s, mol/m³) minus the bulk solution concentration (c, mol/m³) divided by the thickness of the diffusion boundary layer (δ_D, m). D_i (m²/sec) is the diffusion coefficient and the D/δ_D term in Eq. (7.42) is recast as a mass transfer coefficient, k_D (m/sec).

Table 7.3. *Models used to estimate* k_D *for commonly encountered liquid/solid interface geometries adapted from Table 8.3–3 in Cussler (2009). D (m^2/sec) is the diffusion coefficient; q (m/sec) is the Darcy velocity of the fluid; and ν (m^2/sec) is the kinematic viscosity.*

Interface	Equation	Notes
Laminar flow along flat plate	$k_D = 0.646 \left(\dfrac{D}{L}\right)\left(\dfrac{Lq}{\nu}\right)^{1/2}\left(\dfrac{\nu}{D}\right)^{1/3}$	L = plate length
Packed bed	$k_D = 1.17 q \left(\dfrac{Lq}{\nu}\right)^{-0.42}\left(\dfrac{\nu}{D}\right)^{2/3}$	L = grain diameter
Turbulent flow through circular pipe	$k_D = 0.026 \left(\dfrac{D}{L}\right)\left(\dfrac{Lq}{\nu}\right)^{0.8}\left(\dfrac{\nu}{D}\right)^{1/3}$	L = pipe diameter
Laminar flow through circular pipe	$k_D = 1.62 \left(\dfrac{D}{L}\right)\left(\dfrac{L^2 q}{L_p D}\right)^{0.8}$	L = pipe diameter L_p = pipe length
Forced convection around a sphere	$k_D = \left(\dfrac{D}{L}\right)\left[2.0 + 0.6\left(\dfrac{Lq}{\nu}\right)^{1/2}\left(\dfrac{\nu}{D}\right)^{1/3}\right]$	L = sphere diameter
Free convection around a sphere	$k_D = \left(\dfrac{D}{L}\right)\left[2.0 + 0.6\left(\dfrac{L^3 g \Delta \rho}{\rho \nu^2}\right)^{1/4}\left(\dfrac{\nu}{D}\right)^{1/3}\right]$	L = sphere diameter g = gravitational acceleration
Turbulent flow through a horizontal slit	$k_D = 0.026 \left(\dfrac{D}{L}\right)\left(\dfrac{Lq}{\nu}\right)^{0.8}\left(\dfrac{\nu}{D}\right)^{1/3}$	L = ($2/\pi$)(slit width)
Transport through a membrane	$k_D = \left(\dfrac{D}{L}\right)$	L = membrane thickness
Spinning disk	$k_D = 0.62 \left(\dfrac{D}{L}\right)\left(\dfrac{L^2 \omega}{\nu}\right)^{1/2}\left(\dfrac{\nu}{D}\right)^{1/3}$	L = disk diameter ω = rotation rate (radians/sec)

$$J_D = k_D (c_s - c) \tag{7.43}$$

The flux of material into or out of the boundary layer due to reactions at the surface (J_R, mol/m^2sec) is driven by the difference between the surface concentration (c_s, mol/m^3) and the equilibrium concentration (c_{eq}, mol/m^3) and is proportional to the dissolution rate constant (k_+, mol/m^2sec).

$$J_R = k_+ \left(\dfrac{c_{eq} - c_s}{c_{eq}}\right) \tag{7.44}$$

The right-hand side of this relationship is equivalent to $k_+(1 - Q/K)$, so this equation is valid for both dissolution and precipitation at near-equilibrium conditions. Equation (7.44) can also be written in terms of a reaction mass transfer coefficient (k_R, m/sec).

$$J_R = \frac{k_+}{c_{eq}}\left(c_{eq} - c_s\right) \tag{7.45}$$

$$J_R = k_R\left(c_{eq} - c_s\right) \tag{7.46}$$

When a solid is immersed in a solution it will eventually reach a steady-state dissolution rate where the rate of reaction at the surface equals the rate of transport through the static film; $J_D = J_R$. This means that Eq. (7.43) can be set equal to Eq. (7.46).

$$k_R\left(c_{eq} - c_s\right) = k_D\left(c_s - c\right) \tag{7.47}$$

The surface concentration can be found by rearranging Eq. (7.47).

$$c_s = \frac{k_R c_{eq} + k_D c}{k_R + k_D} \tag{7.48}$$

This definition of surface concentration can be substituted into either Eq. (7.43) or Eq. (7.45) to find the combined diffusion and reaction flux (J, mol/m²sec) at steady state.

$$J = \frac{k_R k_D}{k_R + k_D}\left(c_{eq} - c\right) \tag{7.49}$$

According to this equation, if $k_R \gg k_D$, then $k_R + k_D \approx k_R$ so $J = k_D(c_{eq} - c)$, which means that the flux to or from the solution is controlled by diffusion. If $k_D \gg k_R$, then $k_R + k_D \approx k_R$ so $J = k_s(c_{eq} - c)$ and the flux to the solution is controlled by the surface reaction rate.

Values of k_D are influenced by the interface geometry and the fluid flow velocity. Table 7.3 lists some k_D models for geometries that might be useful to geochemists. The estimated accuracy of most of these models is on the order of ±10% but much larger uncertainties are possible (Cussler, 2009).

Example 7.4. Gypsum dissolution as a function of fluid velocity

At Earth surface conditions, gypsum ($CaSO_4 \cdot 2H_2O$) dissolution rates are controlled by a combination of dissolution rate and mass transfer (see Example 7.3). The relative control by each of these processes depends upon the fluid flow velocity. Equation (7.49) can be used to model the overall rate

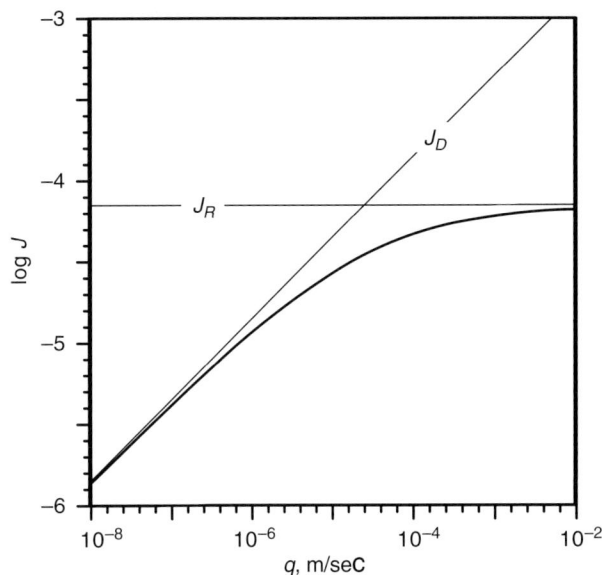

Figure 7.6. Dissolution flux of gypsum into pure water flowing past a flat plate. The line labeled J_R is the dissolution flux due to surface reaction only and the line labeled J_D is for the diffusion flux only.

in terms of the combined effects of dissolution and transport control. This model assumes laminar flow past a flat plate with $L = 0.001$ m and $q = 0.0001$ to 0.1 m/sec.

The data needed for this model are given below.

$c = 0$ mol/m³
$c_{eq} = 15$ mol/m³ (Dutrizac, 2002)
$k_+ = 7.0 \times 10^{-5}$ mol/m²sec (Colombani, 2008)
so $k_R = k_+/c_{eq} = 4.67 \times 10^{-6}$ m/sec
$D_{CaSO_4} = 9.54 \times 10^{-9}$ m²/sec (Example 7.2)
$v = 8.94 \times 10^{-7}$ m²/sec (Watson et al., 1980) ($\rho = 1000$ kg/m³)

$$k_D = 0.646 \left(\frac{D_i}{L}\right)\left(\frac{Lq}{v}\right)^{1/2}\left(\frac{v}{D_i}\right)^{1/3}$$

$$k_D = 0.646 \left(\frac{9.45 \times 10^{-9}}{1 \times 10^{-3}}\right)\left(\frac{1 \times 10^{-3} q}{8.94 \times 10^{-7}}\right)^{1/2}\left(\frac{8.94 \times 10^{-7}}{9.45 \times 10^{-9}}\right)^{1/3} \quad (7.50)$$

$$k_D = 9.52 \times 10^{-4} q^{1/2}$$

Inserting these values into Eq. (7.49) produces the results shown in Figure 7.6. It is relatively easy to appreciate the mixed kinetics of this example. When $k_R \gg k_D$, $k_R + k_D \approx k_D$ and $J = k_D c_{eq}$ and when $k_D \gg k_R$, $k_s + k_D \approx k_s$ and $J = k_R c_{eq}$.

Effect of temperature on mixed kinetics

The diffusion coefficient, the dynamic viscosity, and the mineral dissolution rate all change with temperature. Mineral dissolution reactions have higher activation energies than diffusion processes. This means that increasing the temperature increases the mineral dissolution rate more than the diffusion flux so that mineral dissolution and precipitation processes become limited by the relatively slower diffusion fluxes at high temperatures.

Example 7.5. Gypsum dissolution flux as a function of temperature

Modeling the effect of temperature on gypsum dissolution requires temperature functions for the variables in Eq. (7.49).

$$c_{eq} = 14.4 + 5.88 \times 10^{-2} t - 8.84 \times 10^{-4} t^2 \qquad (7.51)$$

where t = temperature, °C, Dutrizac (2002).

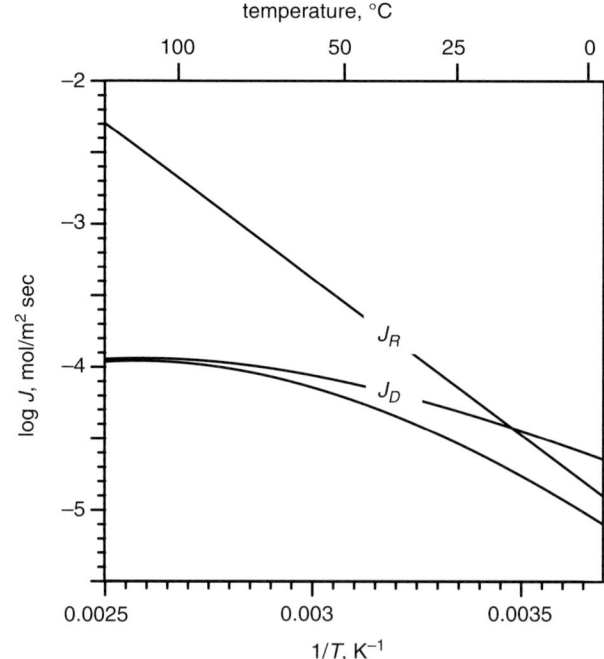

Figure 7.7. Comparison of the $CaSO_4$ fluxes predicted for reaction limited dissolution, $k_R c_{eq}$, transport limited dissolution, $k_D c_{eq}$, and mixed kinetics $(k_R k_D / k_R k_D) c_{eq}$.

The 25°C dissolution rate constant, 7×10^{-5} mol/m² sec (Colombani, 2008), can be extrapolated to other temperatures using the Arrhenius equation and an activation energy of 41.8 kJ/mol (Liu and Nancollas, 1971).

$$\log k_{T_2} = \log k_{T_1} + \frac{E_a}{2.303R}\left(\frac{1}{T_1} - \frac{1}{T_2}\right) = -4.15 + 7.32 - \frac{2183}{T_2} = 3.17 - \frac{2183}{T_2} \quad (7.52)$$

The diffusion coefficient for $CaSO_4$ (D_{CaSO_4}) as a function of temperature is the harmonic mean of D_{Ca} and D_{SO_4} calculated from the temperature functions in Table 7.1.

The kinematic viscosity of water (ν) is calculated from the temperature function for the dynamic viscosity given by Eq. (7.1) and the temperature function for the density given by Eq. (7.2).

This model assumes laminar flow past a flat plate with $L = 0.001$ m and $q = 0.001$ m/sec (Table 7.3). Figure 7.7 shows that the rate of gypsum dissolution changes from reaction limited at $T < 0°C$ to transport limited for $T > 100°$ C. The apparent activation energy for the mixed kinetics model ranges from ~22 kJ/mol at 0°C to ~5 kJ/mol at 100°C.

Examples 7.4 and 7.5 show that diffusive transport tends to become a limiting process for geochemical reactions as flow rates decrease and temperatures increase. This means that although chemical reaction rates often limit geochemical processes at or near the Earth's surface, diffusional transport becomes more rate limiting at depth because of higher temperatures and slower flow rates due to lower rock permeability.

Coating growth model

The rate of interaction between minerals and solution species is sometimes limited by the diffusion of a reactant from the solution through a coating of insoluble reaction products (Figure 7.8). The diffusion rate through this coating can be many orders of magnitude lower than the solution diffusion rate, so that the rate of delivery of the reactant to the mineral surface is much slower than its rate of reaction with the mineral. The thickness of the coating (x, m) is a function of the number of moles of coating precipitated per square meter (n_p, mol/m²), the molar volume of the coating (V_m, m³/mol), the surface area of the grains (A, m²), and the porosity of the coating (ϕ). The coating is likely to be porous if the molar volume of the coating is less than the molar volume of the mineral or if some of the dissolving constituents are lost to the solution. This model is only applicable when x is much smaller than the grain diameter.

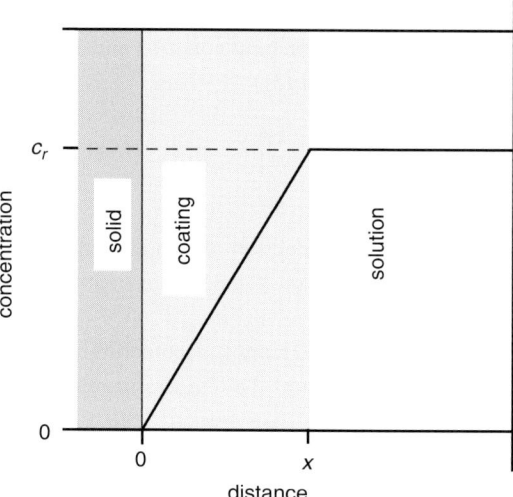

Figure 7.8. Conceptual model showing how the rate of reaction between a mineral with a coating of reaction products is controlled by the rate of diffusion of a reactant from the surrounding solution through the coating to the surface.

$$x = \frac{n_p V_m}{(1-\phi)} \qquad (7.53)$$

The flux of reactant, J_r (mol/m²sec), through the coating is a function of the diffusion coefficient of the reactant in the coating, D (m²/sec), and its concentration in the surrounding solution, c_r (mol/m³).

$$J_r = \frac{D}{x} c_r \qquad (7.54)$$

Although the reactant concentration at the mineral's surface is finite, we will assume that it is so small relative to the solution concentration that it can be set to zero. The flux of constituents from the mineral into the coating is $J_p = v_r J_r$ (mol/m²sec), where v_r = moles of coating formed per moles of mineral dissolved. Combining this relationship with Eqs (7.53) and (7.54) gives an expression for the rate of precipitate formation.

$$J_p = \left(\frac{v_r D c_r (1-\phi)}{V_m} \right) \frac{1}{n_p} \qquad (7.55)$$

The terms in the parentheses can be combined into a rate constant (k_p, mol²/m⁴sec).

$$k_p = \frac{v_r D c_r (1-\phi)}{V_m} \qquad (7.56)$$

Equation (7.55) can be rewritten in terms of k_p and n_p.

$$\frac{dn_p}{dt} = \frac{k_p}{n_p} \qquad (7.57)$$

This equation can be integrated with the boundary conditions that $n_p = 0$ when $t = 0$ and $n_p = n_p$ when $t = t$.

$$\int_0^{n_p} n_p dn_p = k_p \int_0^t dt \qquad (7.58)$$

$$\frac{n_p^2}{2} = k_p t \qquad (7.59)$$

This equation can be solved for n_p.

$$n_p = \left(\sqrt{2k_p}\right) t^{1/2} \qquad (7.60)$$

The thickness of the coating increases with the square root of time.

$$x = V_m n_p = V_m \left(\sqrt{2k_p}\right) t^{1/2} \qquad (7.61)$$

The rate of coating formation is the time derivative of Eq. (7.60).

$$\frac{dn_p}{dt} = \frac{1}{2}\left(\sqrt{2k_p}\right) t^{-1/2} = \left(\sqrt{\frac{k_p}{2}}\right) t^{-1/2} \qquad (7.62)$$

This means that the thickness of the coating increases as a linear function of $t^{-1/2}$.

$$\frac{dx}{dt} = V_m \left(\sqrt{\frac{k_p}{2}}\right) t^{-1/2} \qquad (7.63)$$

As the coating grows, the rate of consumption of the reactant declines as a linear function of the square root of time.

$$\frac{dn_r}{dt} = \left(\frac{1}{v_r}\right)\left(\sqrt{\frac{k_p}{2}}\right) t^{-1/2} \qquad (7.64)$$

This means that a graph of the rate of reactant consumption versus $t^{-1/2}$ is a straight line with a slope of $(1/v_r)(k_p/2)^{1/2}$. The diffusion coefficient for the reactant in the coating can be calculated from this slope by rearranging Eq.

(7.56). This equation is only valid after enough time has passed to allow a sufficiently thick coating to develop so that the flux of the reactant to the mineral surface is smaller than its rate of consumption.

> **Example 7.6.** Iron oxyhydroxide coating on pyrite for AMD mitigation
>
> Huminicki and Rimstidt (2009) suggested that treating pyritic mine wastes with bicarbonate solutions would produce an iron oxyhydroxide coating on the pyrite surfaces, which would inhibit the development of acid mine drainage. The overall reaction converts iron from the pyrite into an iron oxyhydroxide coating.
>
> FeS_2(pyrite) + 3.75 O_2 + 3.5 H_2O + 4 HCO_3^-
> = $Fe(OH)_3$ (iron oxyhydroxide) + 4 H_2CO_3 + 2 SO_4^{2-}
>
> The iron oxyhydroxide coating acts as a barrier to the transport of dissolved oxygen from the solution to the pyrite surface and thereby reduces the rate of pyrite oxidation. They report, based on the experiments of Nicholson *et al.* (1990), that the dissolved oxygen diffusion coefficient through the iron oxyhydroxide coating is ~2×10^{-17} m²/sec, which is about 8 orders of magnitude slower than for pure water. Equation (7.61) can be used to predict the coating thickness as a function of time and Eq. (7.64) can be used to find the rate of O_2 consumption as the coating becomes thicker. The following information is needed to compute k_p.
>
> v_r = number of moles iron oxyhydroxide/number of moles of O_2 reacted = 1/3.75 = 0.267
> $D = 2 \times 10^{-17}$ m²/sec (Huminicki and Rimstidt, 2009)
> $\phi = 0.10$ (assuming that the coating has a porosity of 10%)
> $V_m = 2.72 \times 10^{-5}$ m³/mol
> $c_r = 0.27$ mol/m³ (dissolved O_2 concentration in air-saturated water)
>
> $$k_p = \frac{v_r D c_r (1-\phi)}{V_m} = \frac{(0.267)(2 \times 10^{-17})(0.27)(0.9)}{2.72 \times 10^{-5}} = 4.77 \times 10^{-14} \frac{\text{mol}^2}{\text{m}^4 \text{sec}}$$
>
> The coating thickness as a function of time shown in Figure 7.9a is calculated using Eq. (7.61).
>
> $$x = V_m \left(\sqrt{2k_p}\right) t^{1/2} = (2.72 \times 10^{-5})\left(\sqrt{2(4.77 \times 10^{-14})}\right) t^{1/2} = 8.40 \times 10^{-12} t^{1/2}$$
>
> The oxygen consumption rate as a function of time shown in Figure 7.9b is calculated using Eq. (7.64).
>
> $$\frac{dn_r}{dt} = \left(\frac{1}{v_r}\right)\sqrt{\frac{k_p}{2}} \, t^{-1/2} = (0.267)\left(\sqrt{\frac{4.77 \times 10^{-14}}{2}}\right) t^{1/2} = 4.12 \times 10^{-8} t^{-1/2}$$

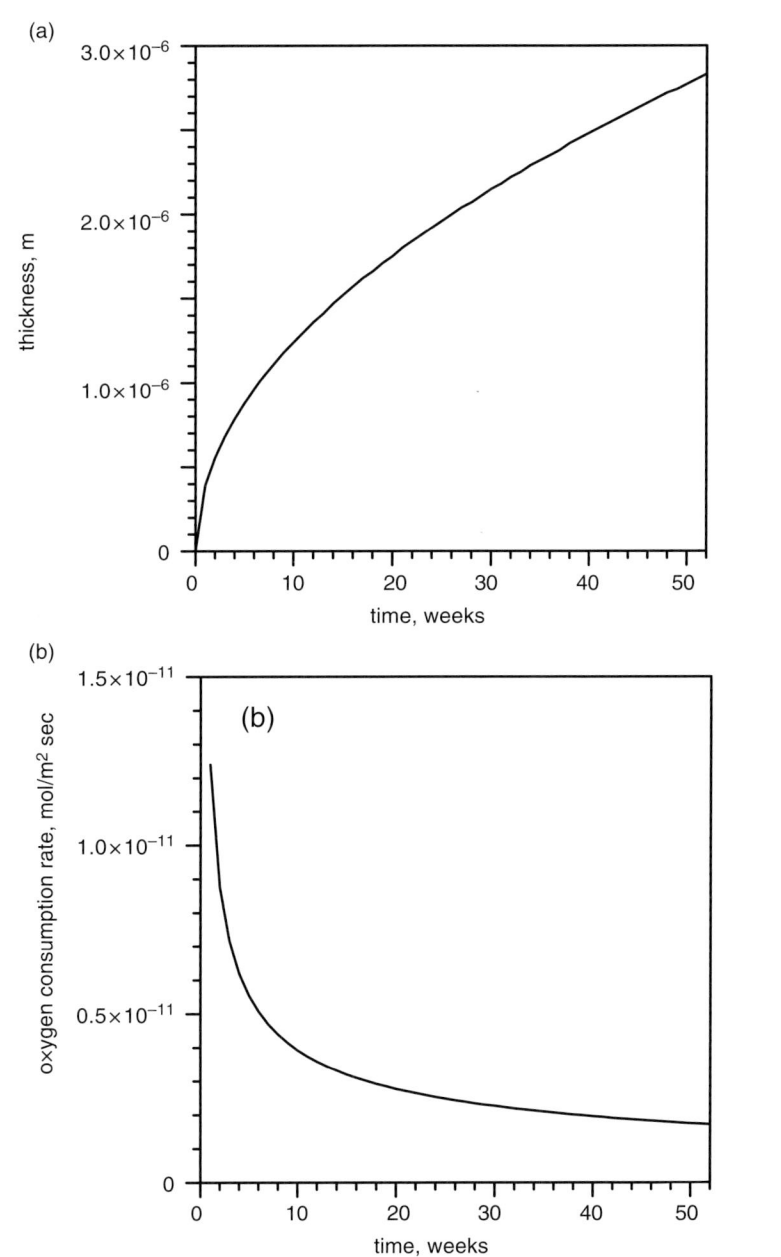

Figure 7.9. (a) Thickness of an iron oxyhydroxide coating on pyrite undergoing oxidation in a solution buffered to a near-neutral pH. (b) Rate of oxygen consumption by the pyrite oxidation reaction as a coating of iron oxyhydroxide grows in thickness.

> The rate of dissolved oxygen consumption by pyrite oxidation in air-saturated water at pH = 7 is 4.80×10^{-10} mol/m²sec based on the rate equation from Williamson and Rimstidt (1994). The model predicts that oxygen flux through the coating after 1 week, 8.83×10^{-13} mol/m²sec, is only about 0.2% of this rate.

Shrinking core models

As the coating on a reacting particle grows thicker, the diameter of the unreacted core decreases and the resulting decrease in the interfacial area between the core and the coating further reduces the rate of delivery of reactant. Figure 7.10 illustrates the geometry of a particle with a shrinking core. The rate of conversion of the core into coating (dn/dt, mol/sec) is a function of the radius of the core (R, m), the stoichiometric coefficient (Δv_i = moles coating formed per moles of core destroyed), the diffusion coefficient of the reactant through the coating (D, m²/sec), and the concentration of the reactant in the surrounding solution (c, mol/m³).

$$\frac{dn}{dt} = -\frac{4\pi R^2}{v_r} D \left(\frac{dc}{dR} \right) \tag{7.65}$$

If the diffusion rate of reactant through the coating is fast compared to the motion of the core/coating interface, a steady-state concentration gradient will develop so that Eq. (7.64) can be integrated across R to find the rate of reaction in terms of R and R_o.

$$\frac{dn}{dt} = \frac{4\pi Dc}{v_r} \left(\frac{RR_o}{R_o - R} \right) \tag{7.66}$$

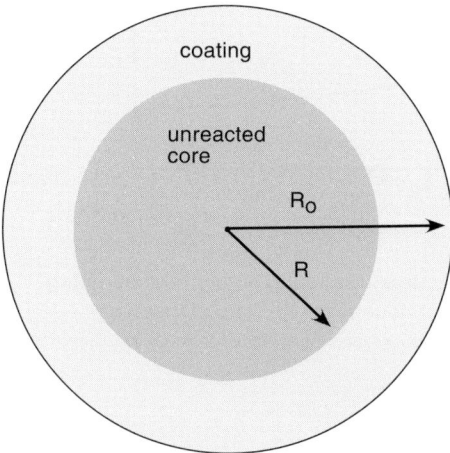

Figure 7.10. Schematic diagram showing the conceptual basis for the shrinking core model.

The number of moles of substance in the particle is $n = 4\pi R^3/3V_m$. The fraction reacted (α) is related to the ratio of the radii of the core and particle cubed.

$$\alpha = 1 - \frac{R^3}{R_o^3} \quad (7.67)$$

This equation can be differentiated with respect to time.

$$\frac{d\alpha}{dt} = \frac{3R^2}{R_o^3}\left(\frac{dR}{dt}\right) \quad (7.68)$$

Substituting Eqs (7.67) and (7.68) into Eq. (7.66) gives the rate of transformation in terms of the fraction of the particle reacted.

$$\frac{d\alpha}{dt} = \frac{3V_m Dc}{v_r R_o^2} \frac{(1-\alpha^{1/3})}{1-(1-\alpha^{1/3})} \quad (7.69)$$

This equation can be integrated using the boundary conditions that $\alpha = 0$ when $t = 0$.

$$1 - \frac{2}{3}\alpha - (1-\alpha)^{2/3} = \left(\frac{2V_m Dc}{v_r R_o^2}\right)t = k_p t \quad \left(k_p = \frac{2V_m Dc}{v_r R_o^2}\right) \quad (7.70)$$

A graph of $1 - \frac{2}{3}\alpha - (1-\alpha)^{2/3}$ versus t has a slope of k_p. Typically k_p is found from experimental data and then used to find D. There are other shrinking core models derived for different geometries and with different boundary conditions (Heizmann et al., 1986; Wen, 1968).

The complexity of Eq. (7.70) creates difficulties for using it in forward models to find α as a function of t. It is easier to graph t as a function of α and then use the graph to find α at the time of interest.

Example 7.7. Development of a limonite pseudomorph after pyrite

Pyrite crystals that oxidize in an environment with a near-neutral pH and with enough alkalinity to neutralize the acid produced by the oxidation reaction typically become replaced by a mixture of iron oxyhydroxide minerals, mostly goethite (FeOOH). This mixture is sometimes referred to as "limonite".

$$FeS_2(\text{pyrite}) + 3.75\ O_2 + 2.5\ H_2O + 4\ HCO_3^-$$
$$= FeOOH(\text{goethite}) + 4\ H_2CO_3 + 2\ SO_4^{2-}$$

The rate of conversion of pyrite to limonite is controlled by the diffusive rate of transfer of dissolved oxygen through the limonite coating. The fraction of pyrite

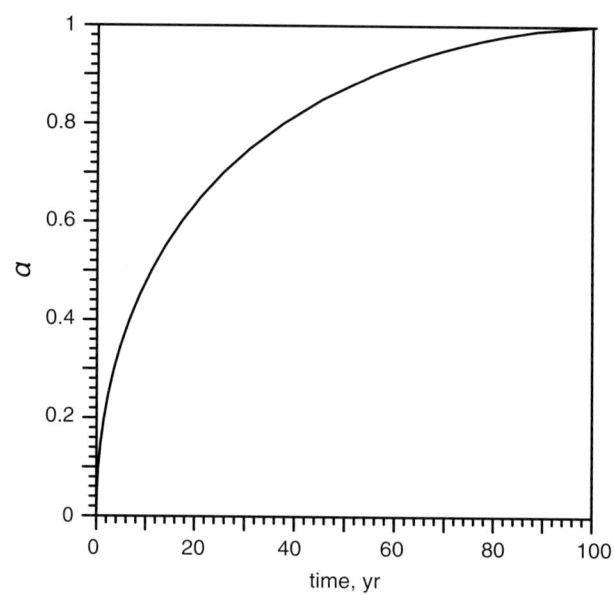

Figure 7.11. Fraction of pyrite converted to goethite as a function of time for a 1 cm diameter (R_o = 0.5 cm) pyrite grain.

converted to goethite, α, can be modeled as a function of time using Eq. (7.70) with the following information.

$V_m(py) = 2.08 \times 10^{-5}$ m³/mol; $V_m(goe) = 2.39 \times 10^{-5}$ m³/mol

$$\phi = 1 - \frac{V_m(goe)}{V_m(py)} = 1 - \frac{2.08 \times 10^{-5}}{2.39 \times 10^{-5}} = 0.13$$

$$\tau^2 = 1 - 2.02 \ln \phi = 1 - 2.02 \ln(0.13) = 9.24$$

$$D_{O_2} = 2.3 \times 10^{-5} \, m^2/sec \text{ (Lerman, 1979)}$$

$$D_{\tau O_2} = \frac{D_{O_2}}{\tau^2} = \frac{2.3 \times 10^{-9}}{9.24} = 2.49 \times 10^{-10} \, m^2/sec$$

$$c_r = 0.27 \, mol/m^3$$

(c_r = dissolved O_2 concentration in air-saturated water)

$$v_r = \frac{\text{moles goethite formed}}{\text{moles } O_2 \text{ reacted}} = \frac{1}{3.75} = 0.267$$

$$R_o = 0.005 \, m = 0.5 \, cm$$

$$k_p = \frac{2V_m Dc}{v_r R_o^2} = \frac{2(2.08 \times 10^{-5})(2.49 \times 10^{-10})(0.27)}{(0.267)(0.01)^2} = 1.05 \times 10^{-10} \text{ sec}^{-1}$$

Using these values we can find the time needed to convert various fractions of a pyrite grain into the goethite coating (Figure 7.11).

$$t = \frac{1 - \frac{2}{3}\alpha - (1-\alpha)^{2/3}}{k_p} = \frac{1 - \frac{2}{3}\alpha - (1-\alpha)^{2/3}}{1.05 \times 10^{-10}}$$

This model shows that the rate of conversion of pyrite to goethite is quite fast during the initial stages of the process so that 50% of the pyrite is converted during the first 11 years of exposure to oxidizing conditions. As the layer of goethite grows thicker, the diffusion distance becomes long and the rate of conversion slows so that it takes an additional 89 years to convert the remaining pyrite to goethite.

Chapter 8
Quasi-kinetics

Quasi-kinetic models deal with processes that are controlled by mass transfer rates rather than by chemical reaction rates. These models assume nearly instantaneous attainment of equilibrium within the region of interest, so changes in the species distribution are controlled by the rate of transfer of substances into or out of that region. These models are constrained by continuity equations making them similar to the chemical reactors models in Chapter 4.

Local equilibrium assumption

Most of the models considered in this chapter rely on the local equilibrium assumption. This assumption requires that the rates of chemical reaction and local mass transfer within the model's spatial domain are fast relative to the residence time of a slug of solution within that domain. Knapp (1989) and Bahr and Rubin (1987) have evaluated conditions where the local equilibrium assumption is valid and the Knapp treatment is summarized by Zhu and Anderson (2002).

For the local equilibrium assumption to be valid, both the mineral and solution reaction rates *and* the transport rate to and from the minerals' surfaces must be fast. These constraints are best tested using the first Damköhler number, Da_I, and the Péclet number, Pe. Da_I compares the rate of consumption (or production) of a species by chemical reaction to the rate of delivery (or removal) of that species by advection.

$$Da_I = \frac{RL}{m_{eq}v} = \frac{\{(A/M)J\}L}{m_{eq}v} \tag{8.1}$$

R (molal/sec) is the reaction rate and m_{eq} (molal) is the equilibrium concentration of the reactant. v (m/sec) is the surficial velocity of the fluid and L is the characteristic length of the domain. A rule of thumb is that when $Da_I < 0.1$, less than 10% of the aqueous species will react as the solution transits L; and when $Da_I > 10$, more than 90% of the aqueous species will react over this distance. Even if the reaction rate is fast, the rate of reaction between solution and the mineral might still be limited by the rate of diffusive and dispersive transport to or from the minerals' surfaces. The Péclet number, Pe, compares the rate of species transport to or from the minerals' surfaces by diffusion or dispersion to the rate that they are added to or removed from the domain by advection. The Péclet number is the ratio of mass transfer by advection to the rate of mass transfer to or from the minerals' surfaces expressed by the sum of the diffusion coefficient (D_i, m^2/sec) and the longitudinal dispersion coefficient (D_L, m^2/sec).

$$Pe = \frac{vL}{D_i + D_L} \qquad (8.2)$$

When Pe < 50, diffusion and dispersion dominate and aqueous species are delivered to or removed from the surface faster than they are added to or removed from the domain by advection. When Pe > 100, the rate of advection is so fast that the reacting species are transported into or out of the domain before they can migrate to or from minerals' surfaces. Figure 8.1 shows a map of Da and Pe. The local equilibrium assumption is valid in the upper right corner of the map where Da > 10 and Pe > 100 but reactions will be unlikely to reach equilibrium for the other conditions shown because they are either transport or reaction limited, or both.

Example 8.1. Equilibration of dissolved silica with quartz in a fracture

The quartz geothermometer (Fournier, 1977) requires that the dissolved silica concentration in an ascending hydrothermal solution becomes "quenched in" as the fluid approaches the surface. On the other hand, the phase transfer model for the rate of quartz deposition, described in this chapter, requires that the LEA be valid. The conditions for which each of these models is appropriate can be found by calculating values of Da_I and Pe. Figure 8.1 shows the temperature range for the transition from LEA to quenched conditions.

The rate of quartz precipitation, R_p (molal/sec), (Rimstidt and Barnes, 1980) is needed to calculate Da_I.

$$R_p = \frac{A}{M} k_+ \left(1 - \frac{Q}{K}\right) \qquad (8.3)$$

For a fracture, the surface area to mass of solution ratio ($A/M = 2/1000\rho W$) depends on the fracture width (W, m) (see Table 6.1). This model considers a 500 μm wide fracture so (A/M) = 4. To simplify the calculations, the solution density (ρ) is set equal to one over the entire temperature range and the degree of supersaturation (Q/K) is set equal to two, which makes $R_p = -(A/M)k_+$. The

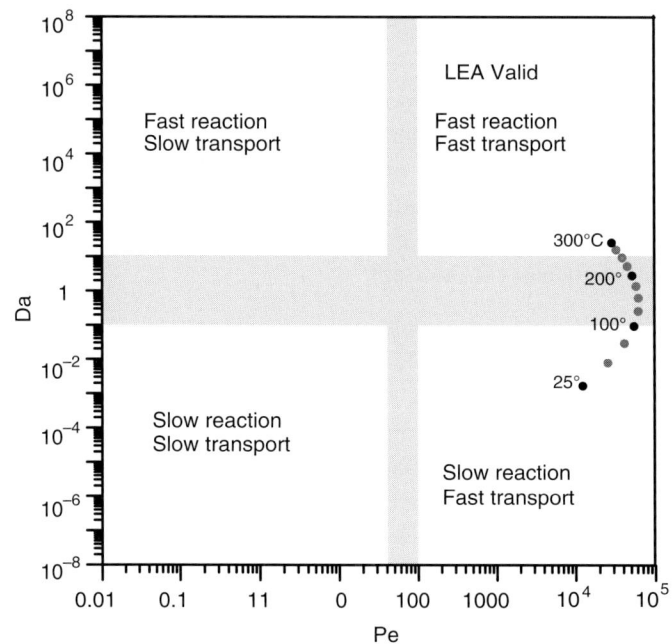

Figure 8.1. Map of Da_I versus Pe that shows that the Local Equilibrium Assumption (LEA) is valid for $Da_I >\sim 10$ and $Pe > \sim 100$. The dots are values of the Da_I and Pe values for quartz precipitation in a fracture at different temperatures calculated in Example 8.1. They show that the LEA is valid for $T > \sim 250°C$.

rate constant, k_+, for each temperature is computed from Rimstidt and Barnes (1980) and Rimstidt (1997b).

$$\log k_+ = -(3705/T) - 0.372 \tag{8.4}$$

$$R_p = (4)(0.185)(10^{-3705/T}) \tag{8.5}$$

The equilibrium quartz concentration, m_{eq}, is calculated from Rimstidt (1997b).

$$\log m = -(1107/T) - 0.025 \tag{8.6}$$

We will consider a characteristic length, L, of 10 m and an ascension rate, v, of 1×10^{-5} m/sec, which is approximately 1 m/day.

It is useful to find Da_I at 25°C to illustrate the calculation procedure.

$$R_p = 2.77 \times 10^{-13}$$

$$L = 10 \text{ m}$$

$$m_{eq} = 1.83 \times 10^{-4} \text{ mol/kg}$$

$$v = 1 \times 10^{-5} \text{ m/sec}$$

$$\text{Da}_\text{I} = \frac{RL}{m_{eq}v} = \frac{(2.77 \times 10^{-13})(10)}{(1.83 \times 10^{-4})(1 \times 10^{-5})} = 1.51 \times 10^{-3} \tag{8.7}$$

Note that if $L = 100$ m, $\text{Da}_\text{I} = 1.5 \times 10^{-2}$, then Da_I scales directly with L. This value of Da_I falls in the "slow reaction" part of Figure 8.1, indicating that the solution does not reach equilibrium within a 10 or even 100 m fracture length because of the very slow precipitation rate of quartz.

The diffusion coefficient for dissolved silica, D_i (m²/sec), and the longitudinal dispersion coefficient, D_L (m²/sec), are needed to compute the Péclet number. D_i is calculated from equation (5) in Rebreanu *et al.* (2008) and D_L is calculated from the fracture width, using the Horne and Rodriguez (1983) model.

$$D_L = \frac{2}{105}\left(\frac{(W/2)^2 v^2}{D_i}\right) \tag{8.8}$$

$$\text{Pe} = \frac{vL}{D_L + D_i} \tag{8.9}$$

Calculating Pe at 25°C requires the following information.

$$L = 10 \text{ m}$$

$$v = 1 \times 10^{-5} \text{ m/sec}$$

$$D_i = 1.1 \times 10^{-9} \text{ m}^2/\text{sec}$$

$$D_L = \frac{2}{105}\left(\frac{(W/2)^2 v^2}{D_i}\right) = \frac{2}{105}\left(\frac{(0.0005/2)^2 (1 \times 10^{-5})^2}{1.1 \times 10^{-9}}\right) = 4.3 \times 10^{-9} \tag{8.10}$$

$$\text{Pe} = \frac{vL}{D_L + D_i} = 2.3 \times 10^4 \tag{8.11}$$

This result means that the transport of silica to the quartz surface by dispersion and diffusion is very fast and is not rate limiting.

Values of Da_I and Pe calculated for other temperatures are plotted on Figure 8.1. The figure shows that the LEA holds for high temperatures (>~250°C) but fails for T <~200°C. This is consistent with using the quartz geothermometer to find the temperature of geothermal reservoirs. The quartz geothermometer is based on the assumption that the dissolved silica concentration remains unchanged as the solution flows to the surface (Fournier, 1977).

Box models

Some models assume that a system reaches a steady state rather than equilibrium. Equilibrium is defined by the principle of detailed balance, which requires that the forward and reverse rates are equal and that each step along the reaction path is reversible. The forward and reverse rates of steady-state processes are equal but the process steps that produce the forward rate are different from those that produce the reverse rate. At steady state, the state variables of an open system remain constant even though there is mass and/or energy flow through the system. The steady-state assumption is especially useful for processes that occur in a series, because the concentrations of intermediates that are formed and subsequently destroyed are constant. Perturbation of a steady-state system produces a transient state where the state variables evolve over time and approach a new steady state asymptotically.

Box models are steady-state models that are used to follow the flow of a conserved substance between reservoirs in the geological environment (Lasaga, 1980b; Lasaga and Berner, 1998; Lerman and Wu, 2008). Box models divide the environment into various reservoirs, represented by boxes (Figure 8.2), each of which represents a major part of the overall system. If the model considers cycling of a substance throughout the entire Earth, the reservoirs might be the oceans, the atmosphere, the crust, etc. For example, Figure 8.3 shows the main reservoirs and fluxes for the global water cycle (Berner and Berner, 1987). It is presumed that each of these reservoirs is well mixed, so the concentration of the substance is approximately the same throughout. Material is transferred from one reservoir to another via fluxes (F) of the substance, which are represented by arrows. Simply constructing a diagram shows how a substance flows through the system (Figures 8.2a and 8.3). In addition, box models are used to quantify how the overall system might respond to a perturbation.

The simplest box model consists of one reservoir that is fed by a constant flux (F_0) (Figure 8.2). The flux out of a reservoir is the product of the mass of the substance (M) in the reservoir and a mass transfer constant (k). Box model reservoirs are simply gigantic ideal mixed flow reactors (see Chapter 4) where there is no net generation or consumption of the substance. Over geologic time spans these reactors tend to attain a steady state so that the flux into a reservoir is matched by the flux out, which means that the rate of accumulation in the reservoir is zero and M is constant (M_{ss}). The flux out of the reservoir equals a mass transfer constant (k, yr^{-1}) times the mass of the substance in the reservoir.

$$F_0 = F = kM_{ss} \tag{8.12}$$

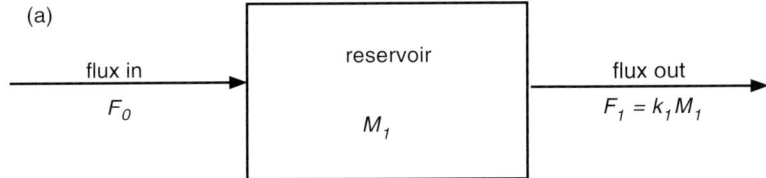

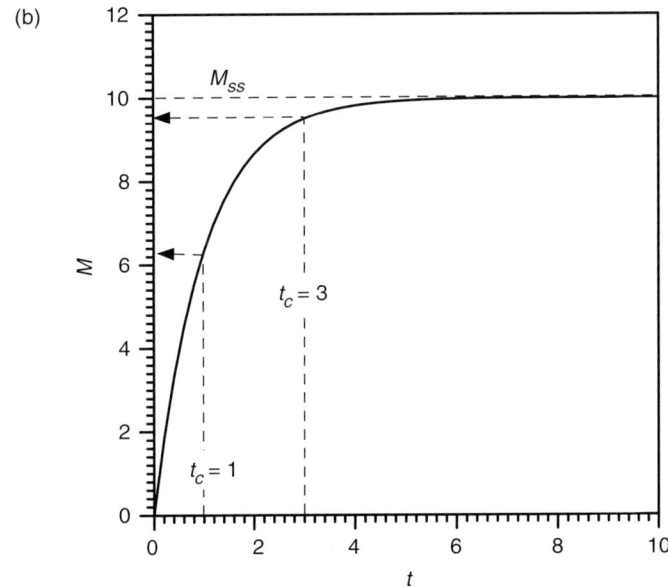

Figure 8.2. (a) Schematic illustration of a box model with a constant flow (F_o) of material into the reservoir, which contains M amount of material. The flux out of the reservoir (F_1) is a function of the mass transfer constant (k_1) and of the amount of material in the reservoir (M_1). (b) Illustration of the transient behavior of a single, initially empty, reservoir as it returns to a steady state M of 10. The reservoir refills to 63% of the steady-state value when one time constant ($t_c = 1/k = 1$) has passed.

$$k = \frac{F_0}{M_{ss}} = \frac{F}{M_{ss}} \qquad (8.13)$$

Once the mass transfer constant is known, it is possible to model the effects of a perturbation on the reservoir contents and output. The rate of change of mass of material in the reservoir is the difference between the flux into and the flux out of the reservoir.

$$\frac{dM}{dt} = F_0 - kM \qquad (8.14)$$

This equation can be rearranged and integrated from an initial condition where $M = M_o$ when $t = 0$.

$$dt = \frac{dM}{F_0 - kM} \qquad (8.15)$$

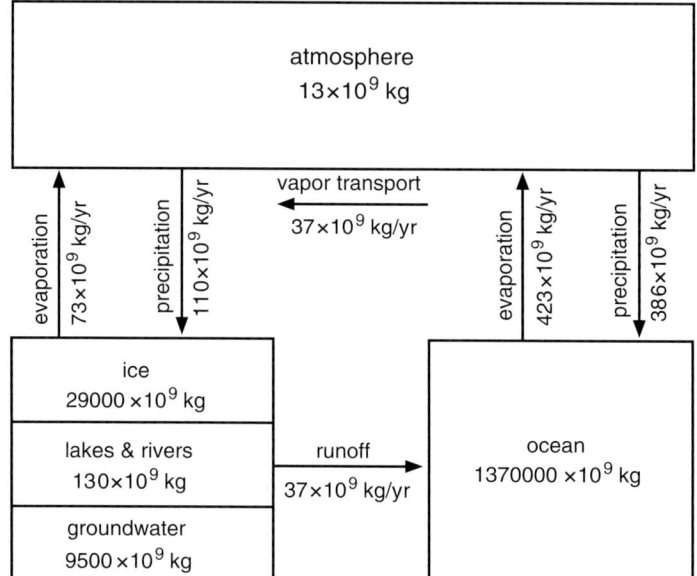

Figure 8.3. Box model of the water cycle (Berner and Berner, 1987) showing the major reservoirs and fluxes.

$$\int_0^t dt = \int_{M_0}^{M} \frac{dM}{F_0 - kM} \tag{8.16}$$

$$t = \left(-\frac{1}{k}\right) \ln\left(\frac{F_0 - kM}{F_0 - kM_0}\right) \tag{8.17}$$

This equation can be solved for M as a function of t.

$$e^{-kt} = \frac{F_0 - kM}{F_0 - kM_0} \tag{8.18}$$

$$M = \frac{F_0}{k} - \left(\frac{F_0}{k} - M_0\right) e^{-kt} \tag{8.19}$$

Figure 8.2b, which is based on Eq. (8.19), shows how an empty reservoir returns to the steady state. During the initial stage of the process the reservoir contains very little of the substance, so the rate of removal by kM is low and M grows quickly. However, increasing M over time causes kM to increase and that causes the rate of growth of M to slow. Eventually, M becomes large enough to cause kM to equal F_0 and M has returned to M_{ss}. Because M approaches M_{ss} as an asymptote, it is not possible to mathematically define a time for the return to steady-state conditions. However, it is possible to find a practical time by defining a time constant (t_c) for the system.

$$t_c = \frac{1}{k} \qquad (8.20)$$

If $1t_c$ is substituted into Eq. (8.19), the calculated value of M is 63% of M_{ss}. At $3t_c$, M is 95% of M_{ss} and at $5t_c$, M is 99% of M_{ss}. Somewhere between $3t_c$ and $5t_c$ the difference between M and M_{ss} becomes smaller than the uncertainty in determining M so that there is no longer a way to distinguish M from M_{ss} and the system has effectively reached steady state.

Although the box models treat substances as a continuum, they actually consist of discrete particles (atoms, molecules, clusters, etc.) and it is useful to consider the average time that a single particle might remain in a reservoir. Some particles will persist in the reservoir for longer times and some for shorter but the average time spent in the reservoir is defined by the residence time (t_{res}).

$$t_{res} = \frac{M_{ss}}{F_0} \qquad (8.21)$$

For a one-reservoir model, $t_{res} = t_c$; see Eq. (8.13). Residence-time analysis is a useful way to understand the behavior of tracers, such as tritiated water, in geochemical cycles. Residence-time analysis methods for more complicated situations are described in many chemical engineering textbooks (Hill Jr., 1977).

Although one-box models are informative, most geochemical cycles consist of multiple reservoirs coupled by many fluxes; see Figure 8.3 for example. Models for multiple reservoir–multiple flux situations and for coupled geochemical cycles are somewhat more complicated but are well understood (Lasaga, 1980a, 1980b; Lasaga and Berner, 1998; Lerman and Wu, 2008). One important insight that comes from these more complex models is that increasing complexity seems to increase the stability of the system. This means that environmental perturbations caused by natural processes or human activities tend to be damped by complex natural systems.

Example 8.2. Perturbation of the global water cycle

Figure 8.3 shows that for the global water cycle, the sum of the water evaporated from the oceans and the continents is 496×10^9 kg/yr. At steady state the precipitation flux must have the same value. The water content of the atmosphere can be modeled as a single reservoir with a water content of 13×10^9 kg, so $k = (496 \times 10^9 \text{ kg/yr}/13 \times 10^9 \text{ kg}) = 38.2$ yr^{-1}. For this simple one-box model, $t_c = t_{res} = 0.026$ yr $= 9.6$ days. This means that the average residence time of a water molecule in the atmosphere is slightly less than 10 days, so we expect that the rainfall flux should respond quickly to a change in the evaporation flux. For example, if

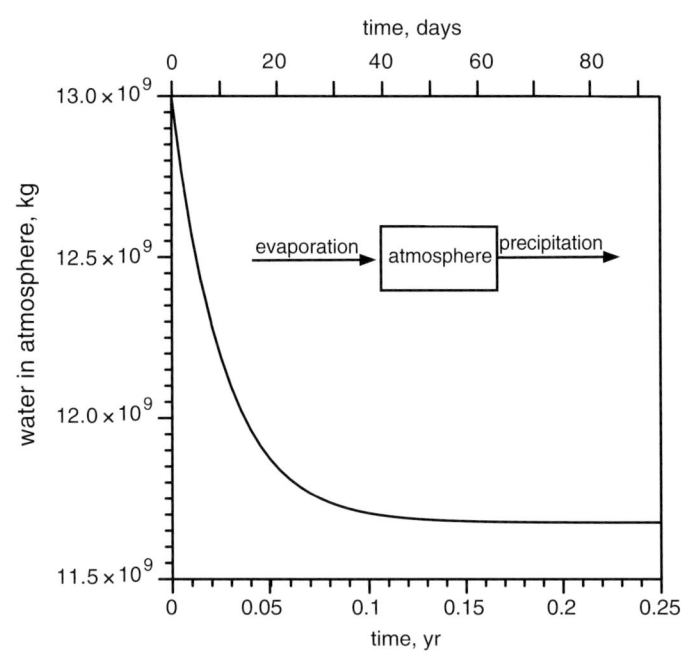

Figure 8.4. Change in the water content of the Earth's atmosphere in response to a sudden 10% decrease in the evaporation rate.

atmospheric particulates from a volcanic eruption were to reduce insolation at the Earth's surface so that the evaporation flux decreases by 10%, F_0 in Eq. (8.19) would drop to 446×10^9 kg/yr. Figure 8.4 shows that this would cause the water content of the atmosphere to decline over a period of about 50 days to a new steady-state value of 11.7×10^9 kg (90% of the previous value).

Phase transfer models

Phase transfer models account for the transfer of material from one phase to another in response to changing intensive or extensive variables. This discussion will be limited to the dissolution or precipitation of minerals in contact with an aqueous solution, but the approach could be used for other scenarios, such as a mineral growing in an igneous melt in response to declining temperature.

Mineral solubility is a function of temperature and pressure (Rimstidt, 1997a) as well as extensive variables such as solution composition. In this model we assume that the extensive variables remain constant and express the solubility of a mineral only in terms of temperature and pressure. The amount of mineral that will dissolve or precipitate in response to changing the temperature and pressure can be found by taking the derivative of the mineral's temperature and pressure solubility function.

$$dm = \left(\frac{\partial m}{\partial T}\right)_P dT + \left(\frac{\partial m}{\partial P}\right)_T dP \qquad (8.22)$$

The partial derivatives in this equation are phase transfer coefficients, α_T and α_P (not to be confused with mass transfer coefficients used in Chapter 7).

$$\alpha_T = \left(\frac{\partial m}{\partial T}\right)_P \text{ and } \alpha_P = \left(\frac{\partial m}{\partial P}\right)_T \qquad (8.23)$$

If a solution is traversing a temperature gradient ($Z'_T = dT/dZ$) and pressure gradient ($Z'_P = dP/dZ$), the change in solubility (dm/dZ, molal/m) along the flow path is the sum of the solubility changes caused by the changing temperature and pressure.

$$\frac{dm}{dZ} = \alpha_T Z'_T + \alpha_P Z'_P \qquad (8.24)$$

If the solution has a Darcy velocity ($q = dZ/dt$) along the direction of the pressure and temperature gradients, the rate of removal of the dissolved constituent from the solution (molal/sec) is

$$R = \frac{dm}{dt} = q(\alpha_T Z'_T + \alpha_P Z'_P) \qquad (8.25)$$

Dividing R by (A/M) gives the flux (J, mol/m²sec) of the dissolved component onto a depositional surface and multiplying J by the molar volume (V_m, m³/mol) of the depositing mineral gives the rate of increase in thickness (dz/dt, m/sec) of a growing layer of that mineral (see Chapter 6).

$$\frac{dz}{dt} = qV_m \left(\frac{M}{A}\right)(\alpha_T Z'_T + \alpha_P Z'_P) \qquad (8.26)$$

Example 8.3. Rate of growth of quartz veins

Most quartz veins are the result of the cooling and decompression of flowing hydrothermal solutions. Because quartz solubility increases with increasing temperature and pressure, the presence of quartz veins in an outcrop indicates that silica-bearing solutions were flowing along a fracture system away from a hot, high-pressure source toward a cooler, lower pressure discharge point.

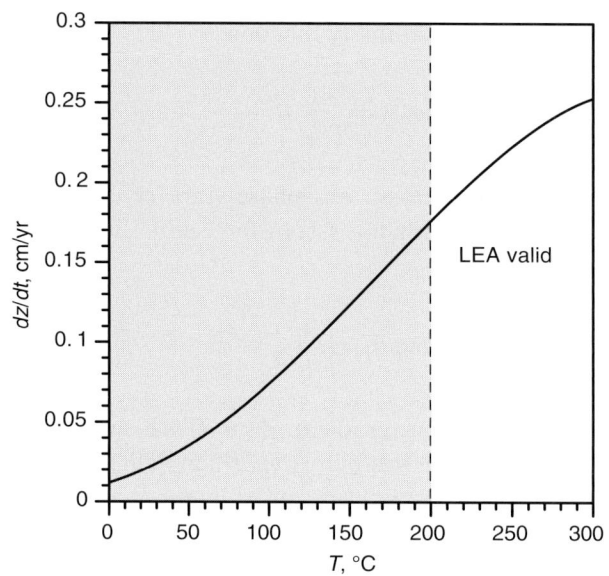

Figure 8.5. Growth rate of a layer of quartz on the wall of a fracture with a 1 cm aperture predicted by the phase transfer model. The quartz is precipitating from an aqueous solution flowing with a Darcy velocity of 1 m/day along a geothermal gradient of 25°/km.

This example estimates the rate of growth of quartz on the wall of a 10 cm wide ($W = 0.1$ m) fracture as the result of the upward flow of a hydrothermal solution at the rate of 4.3 m/day along a geothermal gradient of 25°/km.

$\dfrac{A}{M} = \dfrac{2}{\rho W}$ (Table 6.1) ρ for liquid water is from Wagner and Pruß (2002)

$$q = 4.3 \text{ m/day} = 5 \times 10^{-5} \text{ m/sec}$$

$$Z'_T = 25 \text{ K/km} \,(\sim\text{average geothermal gradient})$$

$$m = 10^{(-1107.12/T - 0.0254)} = e^{(-2549.7/T - 0.0585)} \text{ (Rimstidt, 1997b)}$$

$$\alpha_T = \left(\dfrac{dm}{dT}\right) = \left(\dfrac{2549.7}{T^2}\right)\left(e^{(-2549.7/T - 0.0585)}\right)$$

For relatively shallow earth conditions, we can assume that pressure has little or no effect on solubility so that $\alpha_P = 0$.

When these parameters are inserted into Eq. (8.26) they predict that the rate of growth of quartz on the walls of the fracture (Figure 8.5) ranges from 0.25 cm/yr at 300°C (Z ≈ 11 km) to 0.02 cm/yr at 25°C (surface). This model requires that local equilibrium is maintained (see Example 8.1) so it is only valid for high temperatures ($T > \sim 250°C$). The rates for lower temperatures are too high because the model does not account for the slow precipitation kinetics.

Quartz has prograde solubility so it precipitates as the solution cools. Many minerals (e.g. calcite, barite, gypsum) have retrograde solubility so they precipitate as the temperature increases (Rimstidt, 1997a).

Reaction path models

Geochemical systems typically contain many interacting chemical components distributed among multiple phases, and changing the amount of one or more components reshuffles component distribution among all the phases. Reaction path models predict how the amounts and compositions of phases change in response to the addition or removal of a component or substance from the system. They can be envisioned as a titration experiment in which the extent of reaction, ξ, is advanced by adding or removing a component or substance in a stepwise fashion. The number of moles, n_i, of each substances is linked to the extent of reaction via the reaction coefficient, so that each increment in ξ changes n_i.

$$n_i = n_i^o + v_i \xi \tag{8.27}$$

Reaction path models are local equilibrium models that assume that all the reaction rates are so fast that the system is completely equilibrated at each step. This means that the rates of formation or destruction of all the species in the model are tied by the extent of reaction to the rate of addition, or reaction, of the titrant.

$$\frac{dn_i}{dt} = v_i \frac{d\xi}{dt} \tag{8.28}$$

If the rate of addition of the titrant is specified, then the rates of formation or destruction of all the other phases can be calculated from Eq. (8.28); for example, see Zhu and Lu (2009).

Reaction path models are unreliable whenever the local equilibrium assumption fails. This occurs whenever slow reaction kinetics does not allow a phase to nucleate and grow or when a phase dissolves away too slowly. This means that predictions of reaction path models must be evaluated by a knowledgeable geochemist to identify these refractory phases. The typical remedy for this problem is to remove the uncooperative phase from the model and allow a more reactive, but metastable, phase to take its place.

Most reaction path models are implemented using sophisticated computer programs that can predict the equilibrium redistribution of all species as a substance is added. These computer codes can deal with very large and complex arrays of reactions. The mathematical basis of their algorithms is explained in Helgeson (1968), Helgeson *et al.* (1969), and Helgeson (1979) and their implementation is described by Bethke (2007) and Zhu and Anderson (2002).

Simple reaction path models can be constructed using spreadsheet programs. These models require simplifications such as assuming that the solution is dilute so the activity coefficients are unity and that only the predominant species, which are used to balance the reactions, need be considered in the model.

Example 8.4. Reaction of potassium feldspar with rainwater (from Steinmann *et al.*, 1994)

When rainwater interacts with a rock that contains potassium feldspar, the feldspar dissolves and new minerals grow until the aqueous solution comes to equilibrium with the potassium feldspar. The reactions add K^+ and H_4SiO_4 to the solution and consume H^+. This reaction path model tracks the changing solution composition on an activity diagram (Figure 8.6) to show how the solution composition traverses the stability fields of gibbsite (gib), kaolinite (kaol), and muscovite (mu) until it reaches equilibrium with potassium feldspar (Kf). The phase boundaries are defined by five reactions.

1. kaol + 5 H_2O = 2 gib + 2 H_4SiO_4 $\qquad$ log K_1 = −8.50
2. mu + 9 H_2O + H^+ = gib + K^+ + 3 H_4SiO_4 $\qquad$ log K_2 = −9.38
3. 2 mu + 3 H_2O + 2H^+ = 3 kaol + 2 K^+ $\qquad$ log K_3 = 6.74
4. 3 Kf + 12 H_2O + 2H^+ = mu + 2 K^+ + 6 H_4SiO_4 $\qquad$ log K_4 = −14.41
5. 2 Kf + 8 H_2O + 2H^+ = kaol + 2 K^+ + 4 H_4SiO_4 $\qquad$ log K_5 = −7.36

This model tracks the changes in the concentrations of H^+, K^+, and H_4SiO_4 as small amounts of potassium feldspar are added to 1 kg of rainwater solution. The feldspar first alters to gibbsite, then to kaolinite and then to muscovite. The reaction for each path segment is:

	v_K	v_H	v_{Si}	v_{gib}	v_{kaol}	v_{mu}
A. Kf + H^+ + 7 H_2O = gib + K^+ + 3 H_4SiO_4	1	−1	3	1	0	0
B. Kf + 2 gib + H^+ = 1.5 kaol + K^+ + 0.5 H_2O	1	−1	0	−2	1.5	0
C. Kf + H^+ 4.5 H_2O = 0.5 kaol + K^+ + 2 H_4SiO_4	1	−1	2	0	0.5	0
D. Kf + kaol + 3 H_2O = mu + 2 H_4SiO_4	0	0	2	0	−1	1

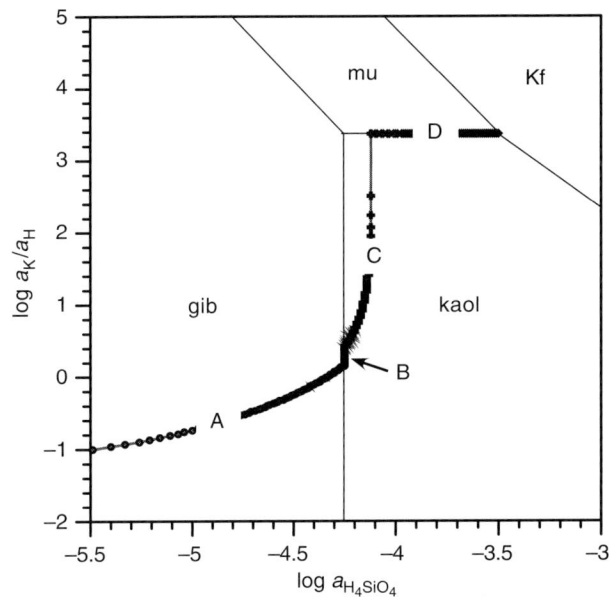

Figure 8.6. Reaction path model of solution composition as potassium feldspar is titrated into rainwater.

The stoichiometric coefficients, v_i, are moles of each species produced or consumed per mole of potassium feldspar reacted.

Slightly acidic rainwater has a H^+ concentration of about 3.2×10^{-5} mol/kg (pH ~4.5) and a K^+ concentration of about 3.2×10^{-6} mol/kg (0.12 ppm) and contains about 5.8×10^{-6} mol/kg (0.1 ppm) dissolved silica.

The reaction path begins at log $a_K/a_H = -1.0$ and log $a_{H4SiO4} = -5.2$. As the amount of K and Si in the system is incremented by the addition of potassium feldspar, reaction A occurs followed by reactions B, C, and D. The stepwise increase in the extent of reaction, ξ, is reflected by changes in the concentration (activity) of the aqueous species and by changes in the number of moles of each mineral.

$$m_{H^+} = \left(m_{H^+}\right)_o + v_{H^+}\xi$$

$$m_{K^+} = \left(m_{K^+}\right)_o + v_{K^+}\xi$$

$$m_{H_4SiO_4} = \left(m_{H_4SiO_4}\right)_o + v_{H_4SiO_4}\xi$$

$$n_{gib} = \left(n_{gib}\right)_o + v_{gib}\xi$$

$$n_{\text{kaol}} = \left(n_{\text{kaol}}\right)_o + \nu_{\text{kaol}}\xi$$

$$n_{\text{mu}} = \left(n_{\text{mu}}\right)_o + \nu_{\text{mu}}\xi$$

The size of the ξ step must be slightly smaller than the change in hydrogen ion concentration at the end of the reaction path otherwise its concentration will become negative. This choice can be made by trial and error.

The resulting diagram shows that the first increments of potassium feldspar are converted to gibbsite as the solution tracks along path A. Then the gibbsite formed along path A dissolves and its components along with those from the added potassium feldspar are converted to kaolinite along path B. After the gibbsite is consumed, potassium feldspar is converted to kaolinite along path C until the solution composition reaches saturation with respect to muscovite. Finally, potassium feldspar is converted to kaolinite and muscovite until the solution composition reaches the muscovite–kaolinite–potassium feldspar triple point. At that point the solution is in equilibrium with potassium feldspar and no further reaction takes place.

Mass balance models

Mass balance models deduce the extent of reaction between a solution and a group of minerals in a batch reactor from the difference in the element concentrations in the initial and final solutions. Mass balance models were developed to interpret mineral weathering reactions (Garrels and MacKenzie, 1967) but the approach is general and the method could be extended to many other situations. Mass balance models are often applied to understand changes in groundwater composition due to reactions that take place along the flow path. In this case, the models must assume that the mineralogy is the same from place to place along the flow path.

The first step in developing a mass balance model is to compute the difference in the concentration of each element between the final and initial solution. In an ideal batch reactor this difference is the result of reactions between the solution and the solid phases. The next step is to postulate reactions that might have caused those changes in the solution composition and to construct a set of linear equations to represent the effect of each reaction. Finally, the linear equations are solved to determine the extent of each of the reactions that must have occurred. If a reasonable solution is not found, the postulated reactions are revised until a fit is found between the reactions and the changes in solution chemistry. This means that the model may not produce a unique fit to the data.

Although mass balance and reaction path models share a conceptual basis, they are quite different in their implementation. Reaction path models begin with an initial solution and rock composition and use equilibrium reactions to predict a final solution and rock mineralogy. Mass balance models use initial and final solution compositions and rock mineralogy to determine the extent of mineral solution reactions. Those reactions need not have reached equilibrium.

Mass balance models can be extended to account for solution speciation and stable isotope exchange but the computations become unwieldy (Plummer et al., 1983). Computer codes are needed to deal with these more complicated situations (El-Kadi et al., 2011; Plummer et al., 1991, 1992).

Example 8.5. Weathering of the Sierra Nevada batholith

Garrels and MacKenzie (1967) published the classic example of the application of a mass balance model to account for chemical weathering in the granitic rocks of the Sierra Nevada batholith. The model, which uses water analyses from Feth et al. (1964), takes the chemical analysis for snowmelt as an initial water composition and the average ephemeral spring water as a final water composition. It is presumed that the infiltrating snowmelt dissolved CO_2 from soil gases to make H_2CO_3 that reacted with plagioclase, potassium feldspar, biotite, and quartz to form kaolinite. The difference between the snowmelt and spring water compositions (Δm) is the result of the reaction of the water with the minerals in the batholith.

Species	Spring water μmol/kg	Snow melt μmol/kg	Δm μmol/kg
Na^+	134	24	110
Ca^{2+}	78	10	68
Mg^{2+}	29	07	22
K	28	08	20
HCO_3^-	328	18	310
$SiO_2(aq)$	273	3	270

The minerals in the batholith that are available for weathering reactions are plagioclase, which consists of albite ($NaAlSi_3O_8$) and anorthite ($CaAl_2Si_2O_8$); biotite ($KMg_3AlSi_3O_{10}(OH)_2$); potassium feldspar ($KAlSi_3O_8$); and quartz (SiO_2). The reactions produce kaolinite ($Al_2Si_2O_5(OH)_4$).

There are five postulated weathering reactions. They are written as a series of linear equations. This is the key feature of the mass balance approach.

(albite–ab) $2\ ab + 9\ H_2O + 2\ H_2CO_3 - kaol - 2\ Na^+ - 2\ HCO_3^- - 4\ H_4SiO_4 = 0$

(anorthite–an) $an + H_2O + 2\ H_2CO_3 - kaol - Ca^{2+} - 2\ HCO_3^- = 0$

(biotite–bio) $2\text{ bio} + H_2O + 14H_2CO_3 - \text{kaol} - 2K^+ - 6Mg^{2+} - 14HCO_3^-$
$- 4H_4SiO_4 = 0$

(K-feldspar–Kf) $2\text{ Kf} + 9 H_2O + 2 H_2CO_3 - \text{kaol} - 2 K^+ - 2 HCO_3^-$
$- 4 H_4SiO_4 = 0$

(quartz–qz) $\text{qz} + 2 H_2O - H_4SiO_4 = 0$

The objective is to determine the extent of each reaction, ξ, by distributing the change in water composition over these five reactions. This can be done by accounting for how much of each aqueous species is consumed or produced by each of the reactions.

species	(ab)	(an)	(bio)	(Kf)	(qz)	Δm, μmol/kg
Na	−2 ab	+ 0 an	+ 0 bio	+ 0 Kf	+ 0 qz	$= 110 = \Delta m_{Na}$
Ca	0 ab	− 1 an	+ 0 bio	+ 0 Kf	+ 0 qz	$= 68 = \Delta m_{Ca}$
Mg	0 ab	− 0 an	− 6 bio	+ 0 Kf	+ 0 qz	$= 22 = \Delta m_{Mg}$
K	0 ab	+ 0 an	− 2 bio	− 2 Kf	+ 0 qz	$= 20 = \Delta m_K$
HCO$_3$	−2 ab	− 2 an	− 14 bio	− 2 Kf	+ 0 qz	$= 310 = \Delta m_{HCO_3}$
H$_4$SiO$_4$	−4 ab	+ 0 an	− 4 bio	− 4 Kf	− 1 qz	$= 270 = \Delta m_{H_4SiO_4}$

There are six equations and five unknowns. The extent of each reaction can be estimated using multiple linear regression. Multiplying the estimated extent of reaction, $\Delta\xi_i$ (values in parentheses are the standard error of the estimate) by the reaction coefficient for the mineral gives the number of moles of that mineral consumed per kilogram of water. If the contact time, Δt, is known, the rate of each reaction can be estimated from $\Delta\xi_i/\Delta t$.

reaction	ξ_i, μmol/kg	v_i	$\Delta\xi_i v_i$, μmol/kg
(ab)	−55(0.5)	2	−110
(an)	−68(0.8)	1	−68
(bio)	−3.7(0.1)	2	−7.3
(Kf)	−6.3(0.6)	2	−12.6
(qz)	−10(3)	2	−20

Garrels and MacKenzie (1967) solved for these values using a sequential subtraction process. The more general multiple linear regression method used here distributes the rounding and analytical errors over the ξ_i estimates and is amenable to expansion to a much larger set of reactions.

Partition coefficients

The trace element concentrations in geologic materials are strongly linked to major element behaviors. This means that trace element distributions offer useful clues about processes that control the distribution and redistribution of major elements. The distribution of a trace element between a solid and solution is expressed by a distribution coefficient (McIntire, 1963; Rimstidt et al., 1998), which is the ratio of the mole fraction of the trace (X_{Tr}) and major (X_M) element in the solid to the ratio of the concentration of the trace (m_{Tr}) to major (m_M) element in the solution.

$$\lambda = \left(\frac{X_{Tr}}{X_M}\right) \bigg/ \left(\frac{m_{Tr}}{m_M}\right) \tag{8.29}$$

This definition incorporates the activity coefficients for the trace element in the solid and solution into λ; see McIntire (1963) for an approach that explicitly accounts for activity coefficients.

The model assumes that during the reaction process the solid's surface is always in equilibrium with the solution. Then it becomes completely isolated when it is overgrown by the next increment of precipitating solid. At low temperatures, the diffusion rate of trace elements in most solids is very slow, so the trace element concentration in each layer remains unchanged once that layer is overgrown.

If $\lambda > 1$, the trace element is enriched in the solid relative to the solution during precipitation, so m_{Tr}/m_M is reduced as precipitation proceeds. If $\lambda < 1$, the trace element is enriched in the solution relative to the solid during precipitation, so m_{Tr}/m_M increases as precipitation progresses. In a mixed flow reactor operating at a steady state, m_{Tr}/m_M remains constant and the solid has a constant X_{Tr}/X_M ratio throughout. If the solid precipitates in a batch reactor, m_{Tr}/m_M changes during the precipitation process causing the X_{Tr}/X_M ratio to change from the center to the surface of the precipitating solid.

Mixed flow reactors

In a mixed flow reactor operating at steady state, the concentrations of Tr and M in the solution are constant. This means that the composition of the solid is also constant. X_{Tr}/X_M reflects the relative rates of removal of the trace and major elements from the solution. These rates are the difference between the concentration of the element in the feed and effluent solutions multiplied by the flow rate (Q, kg/sec) through the reactor.

$$\frac{X_{Tr}}{X_M} = \frac{R_{Tr}}{R_M} = \frac{\left((m_{Tr})_{in} - (m_{Tr})_{out}\right)Q}{\left((m_M)_{in} - (m_M)_{out}\right)Q} = \frac{(m_{Tr})_{in} - (m_{Tr})_{out}}{(m_M)_{in} - (m_M)_{out}} \tag{8.30}$$

Combining Eqs (8.29) and (8.30) and realizing that the effluent solution is a sample of the solution in the reactor leads to

$$\lambda = \frac{(m_{Tr})_{in} - (m_{Tr})_{out}}{(m_M)_{in} - (m_M)_{out}} \bigg/ \frac{(m_{Tr})_{out}}{(m_M)_{out}} = \frac{\left((m_{Tr})_{in}/(m_{Tr})_{out}\right) - 1}{\left((m_M)_{in}/(m_M)_{out}\right) - 1} \quad (8.31)$$

This means that λ can be computed from the difference between the concentration of Tr and M in the feed and effluent solutions regardless of the flow or precipitation rate. These same data along with the flow rate can be used to find the precipitation rate of the major element, so that a mixed flow reactor experiment is well suited to quantify the effect of the precipitation rate on λ.

> **Example 8.6.** Trapping toxic trace elements from solutions using a mixed flow reactor
>
> Trace elements can be removed from dilute solutions by co-precipitation with a major element. For example, a longstanding method used to collect radium from very dilute solution is to co-precipitate it with barium sulfate (Gordon and Rowley, 1957).
>
> Because calcium and bicarbonate are non-toxic and widely available, co-precipitation of radioactive or toxic trace elements with calcite is a reasonable strategy to clean up industrial or radioactive waste solutions (Curti, 1999). Because mixed flow reactors can process a continuous stream of solution they might be employed to remove radioactive Co or radioactive Sr from a solution by co-precipitation with calcite. The reactor would mix a bicarbonate-rich solution with a calcium-rich solution containing the trace element, which would be trapped in the precipitating calcite. Let p ($= m_{out}/m_{in}$) be the fraction of M or Tr that remains in the effluent solution so that α ($= 1 - p$) is the fraction that is precipitated in the reactor. The effectiveness of such a process can be evaluated by rewriting Eq. (8.31) first in terms of p_{Tr} and p_M.
>
> $$\lambda = \frac{(1/p_{Tr}) - 1}{(1/p_M) - 1} \quad (8.32)$$
>
> Equation (8.32) is then written in terms of α_{Tr} and α_M.
>
> $$\lambda = \frac{\left(1/(1-\alpha_{Tr})\right) - 1}{\left(1/(1-\alpha_M)\right) - 1} \quad (8.33)$$
>
> Further manipulation moves α_{Tr} to the left-hand side.
>
> $$\frac{1}{1-\alpha_{Tr}} = \lambda\left(\frac{1}{1-\alpha_M} - 1\right) + 1 \quad (8.34)$$
>
> Taking the reciprocal of both sides of Eq. (8.34) allows it to be solved for α_{Tr}.

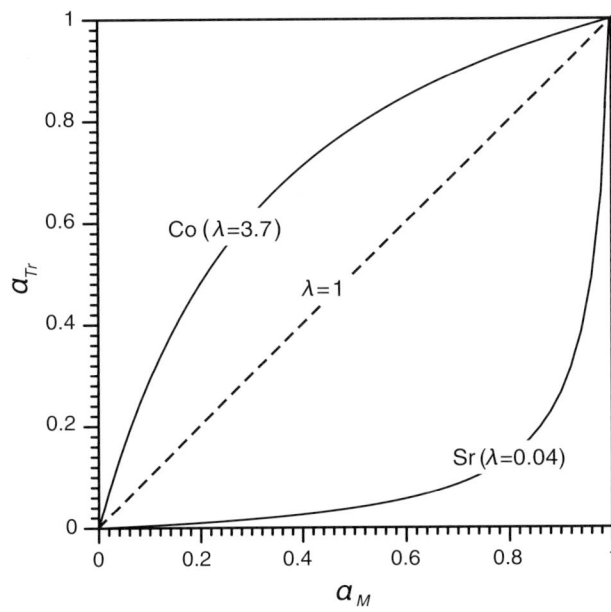

Figure 8.7. Fraction of trace element removed from solution as a function of the fraction of major element precipitated in a mixed flow reactor.

$$\alpha_{Tr} = 1 - \left(1 \Big/ \left(\lambda \left(\frac{1}{1-\alpha_M} - 1 \right) + 1 \right) \right) \qquad (8.35)$$

This equation and λ values from Curti (1999) can be used to model the incorporation of Co and Sr into calcite precipitating in a mixed flow reactor.

$$\lambda_{Co} = 3.7$$
$$\lambda_{Sr} = 0.04$$

Graphs of Eq. (8.35) shown in Figure 8.7 demonstrate that this scheme could work for Co because precipitating 20% of the Ca would remove nearly 50% of the Co. Using a series of mixed flow reactors would further improve the process because removal of 50% of the remaining Co in a second reactor would leave 25% and a third reactor would further reduce the Co to 12.5%. On the other hand, this process would be inefficient for Sr because precipitating 80% of the Ca would only remove about 10% of the Sr. Hence, co-precipitation in a mixed flow reactor is only effective for trapping trace elements if $\lambda > 1$.

Batch reactors

Co-precipitation in batch reactors is modeled using the Doerner–Hoskins approach (Doerner and Hoskins, 1925). Because batch reactors are closed systems, differences in the removal rate of the trace and major elements from the solution cause the solution to become relatively enriched or depleted in the trace element as the precipitation proceeds. This simple model from Curti (1997) considers how removing a small amount of Tr and M by precipitating a small mass of solid affects the composition of the solution.

$$\left(\frac{X_{Tr}}{X_M}\right)_{surface} = \frac{-\Delta m_{Tr}}{-\Delta m_M} = \lambda \left(\frac{m_{Tr} + \Delta m_{Tr}}{m_M + \Delta m_M}\right) \tag{8.36}$$

As the amount of solid precipitated tends to zero, $\Delta m_{Tr} \rightarrow dm_{Tr}$ and $\Delta m_M \rightarrow dm_M$ so

$$\frac{dm_{Tr}}{dm_M} = \lambda \frac{m_{Tr}}{m_M} \tag{8.37}$$

Integrating this relation from initial concentrations gives a relationship that links the concentration of Tr to the concentration of M.

$$\ln\left(\frac{m_{Tr}}{m_{Tr}^o}\right) = \lambda \ln\left(\frac{m_M}{m_M^o}\right) \tag{8.38}$$

This expression allows the calculation of λ from the initial and final concentrations of the trace and major element.

$$\lambda = \ln\left(\frac{m_{Tr}}{m_{Tr}^o}\right) \bigg/ \ln\left(\frac{m_M}{m_M^o}\right) \tag{8.39}$$

Applying the antilog transform to Eq. (8.38) gives a relationship that predicts the concentration of the trace element in the solution as a function of the amount of major element precipitated.

$$\frac{m_{Tr}}{m_{Tr}^o} = \left(\frac{m_M}{m_M^o}\right)^\lambda \tag{8.40}$$

Example 8.7. Trace elements in calcite cements

The concentrations of Fe, Mn, and Mg in calcite cements are often used to interpret their depositional environment. These interpretations use distribution coefficients to estimate the composition of the solution based on an analysis of the calcite cement. Laboratory determinations of these distribution coefficients are typically carried out in batch reactors. Creating a model of the precipitation process in these experiments is helpful for understanding the uncertainties associated with the measurements. If the calcite is precipitated in a batch reactor, Eq. (8.40)

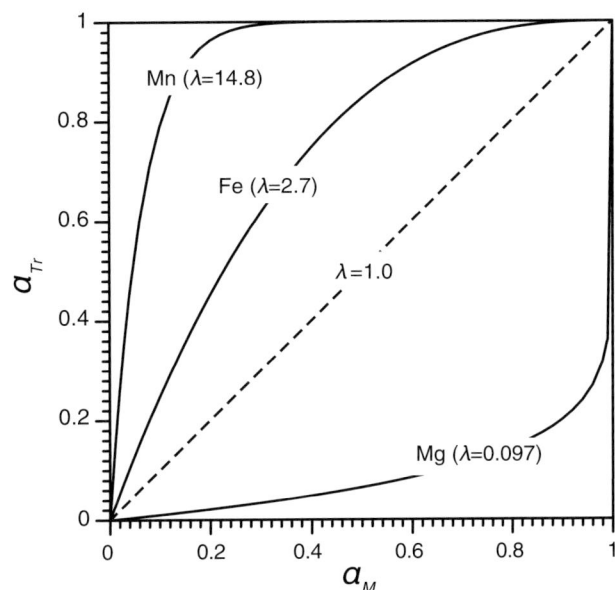

Figure 8.8. Fraction of trace element removed from solution as a function of the fraction of major element precipitated in a batch reactor.

can be modified to model the co-precipitation of the trace elements. Referring to the notation used in Example 8.3, m/m^o is the fraction of element that remains in solution ($= p$) and α ($= 1 - p$) is the fraction that is precipitated.

Equation (8.40) can be recast in terms of the fraction of element remaining in solution.

$$p_{Tr} = (p_M)^\lambda \qquad (8.41)$$

This equation can be further recast in terms of the fraction of the element precipitated.

$$1 - \alpha_{Tr} = (1 - \alpha_M)^\lambda \qquad (8.42)$$

Equation (8.42) can be rearranged to find the fraction of Tr that has precipitated as a function of the fraction of M that has precipitated using the λ values listed below.

$$\alpha_{Tr} = 1 - (1 - \alpha_M)^\lambda \qquad (8.43)$$

$\lambda_{Mn} = 14.8$ (Curti, 1999)

$\lambda_{Fe} = 2.7$ (Curti, 1999)

$$\lambda_{Mg} = 0.097 \text{ (Curti, 1999)}$$

Figure 8.8 shows that within analytical resolution all the Mn is removed from solution by the time 30% of the Ca has precipitated. This means that to determine λ_{Mn}, only small amounts of Ca can be allowed to precipitate and the value of λ_{Mn} will be strongly affected by the Ca analysis. On the other hand, determining λ_{Mg} requires the precipitation of most of the Ca and the accuracy of λ_{Mg} depends upon accurate Mg determinations. Values of λ near 1 are easiest to determine and most accurate. Note that these same constraints apply to determining distribution coefficients using mixed flow reactors (see Figure 8.7).

Chromatography models and retardation factors

Chromatographic separation occurs when the movement of dissolved species is retarded by interaction with a stationary phase along a flow path. Chromatography is a well-established way to separate chemical species in the laboratory. Approaches used to model laboratory chromatography processes are described by Laub (1985), Cazes and Scott (2002), and Arnold *et al*. (1986).

Chromatographic effects occur in natural settings when fluids percolate through and interact with rock or soil materials. There are several one-dimensional models of chromatographic processes for high-temperature processes (Appelo, 1996; Guy, 1993; Hofmann, 1972; Korzhinskii, 1970). All chromatography models assume that a chemical component in the moving fluid equilibrates instantly with a mineral phase as a slug of the fluid moves through the rock, i.e. local equilibrium is maintained. This assumption is combined with a continuity relationship to produce a differential equation that can be used to model the rate of transport of one or more components along the flow path. Chromatography models have been applied to metasomatic infiltration processes such as those that produce skarns. When viewed at a point along the flow path, the moving fluid transports reactants to that point where they react with the minerals until the mineralogy is transformed to an assemblage that is in equilibrium with the fluid. This means that the extent of the reaction and mineralogy at each point along the flow path could be modeled using a reaction path-type model. These infiltration processes tend to produce sharp reaction fronts (Guy, 1993; Helfferich, 1989; Korzhinskii, 1970; Schechter *et al*., 1987) that are related to phase boundary crossings (see Example 8.4). Advancing reaction fronts can interact with each other in two and three dimensions to produce interesting patterns (Dria *et al*., 1987). These chromatography models for infiltration processes at metamorphic conditions are successful because the reaction rates are high and the infiltration rates are low so local equilibrium is maintained.

For near surface conditions, mineral reaction rates slow and the flow rates are relatively faster so the local equilibrium assumption often fails and chromatography models are less useful for modeling mineral transformations. However, many adsorption–desorption processes are fast enough to maintain local equilibrium so that chromatography models are useful for modeling the retardation of species that adsorb to aquifer minerals. The central feature of these models is the retardation factor R_f (no units), which is the ratio of the velocity of transport of the adsorbing species to the velocity of a non-adsorbing, unreactive tracer (Higgins, 1959).

$$R_f = 1 + \frac{\rho_b}{\phi} K_d \tag{8.44}$$

ρ_b (g/cm³) is the bulk density of the packed bed, ϕ (no units) is the porosity, and K_d (cm³/g) is the distribution coefficient. See Zheng and Bennett (1995) for a derivation of this equation. It is related to the K_D (mol/g-solid/mol/kg-water) derived in Chapter 6 by a units transformation involving the molecular weight of the adsorbing species and the density of water. Because K_d is related to K_D, its value is also a function of the specific surface area of the solid.

$$K_D \left(\frac{\text{mol}_i / \text{g}_{\text{solid}}}{\text{mol}_i / \text{kg}_{\text{soln}}} \times \frac{1}{\text{kg}_{\text{soln}} / 1000\ \text{cm}^3_{\text{soln}}} \right) = 1000 K_D = K_d \left(\frac{\text{cm}^3_{\text{soln}}}{\text{g}_{\text{solid}}} \right) \tag{8.45}$$

Example 8.8. Zn^{2+} migration in a lateritic soil

Chotantarat et al. (2011) performed a column experiment using a pH 5 Zn^{2+} solution ($c_o = 4.61 \times 10^{-3}$ mol/L), which was pumped through a 10 cm long column packed with a lateritic soil from the Akara mine area in Thailand. The bulk density of the soil in the packed column was 0.92 g/cm³ and the porosity was 0.66. The columns were calibrated using an inert Br^- tracer (Figure 8.9a), which showed that the breakthrough, where $c/c_o = 0.5$, occurred when 1.0 pore volumes were eluted. The Zn^{2+} experiment (Figure 8.9b) showed that $c/c_o = 0.5$ occurred when 26 pore volumes were eluted so $R_f = 26$. The simplest way to analyze these data is to find K_d from the rearranged Eq. (8.44).

$$K_d = \frac{R_f - 1}{\rho_b / \phi} = \frac{26 - 1}{0.92 / 0.66} = 17.9 \tag{8.46}$$

More sophisticated methods are typically used to find R_f (Nkedl-Kizza et al., 1987) and the adsorption behavior is often modeled using more complex isotherms (Chotantarat et al., 2011) but the principles are the same.

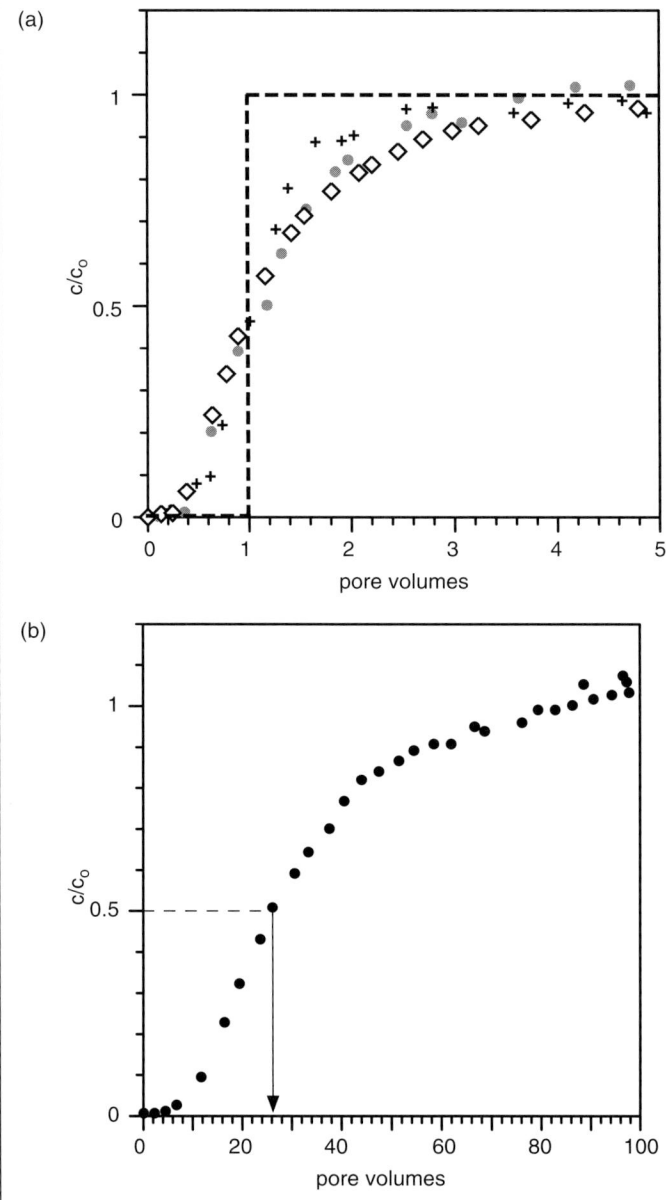

Figure 8.9. (a) Breakthrough curve for three Br⁻ tracer experiments showing that the inert tracer breakthrough occurs at one pore volume. The dashed line shows the pattern expected for ideal plug flow and the deviation of concentration from this line is the result of dispersion and diffusion. (b) Breakthrough curve for the Zn^{2+} experiment showing that c/c_o occurs after 26 pore volumes have eluted.

If a mixture of aqueous species with different K_d values was injected into a plug flow reactor as a spike (small slug), adsorption processes would cause some species to be retarded more than others. This would cause each species to elute from the reactor at a different time. This is the basis for chromatographic separations in the laboratory.

Chapter 9
Accretion and transformation kinetics

Under near-equilibrium conditions, solids form from aqueous solutions via the addition of monomer growth units. The terrace, ledge, and kink (TLK) model and the Burton–Cabrera–Frank (BCF) theory give reasonable descriptions of this process. At higher degrees of supersaturation, monomers addition is joined by the accretion of polymer and larger growth units, up to and including stable crystallites. The solids formed by accretion of these larger units are often metastable and eventually transform to more stable forms. This means that the formation of a stable crystalline solid from a very supersaturated solution involves several steps. This chapter presents models that describe the formation and accretion of larger growth units and transformation of the resulting metastable solids.

Polymerization rates

Polymers such as DNA, proteins, polysaccharides, and polyphenols are essential components of organisms; and synthetic polymers such as Nylon, polyethylene, styrene, and Teflon are basic raw materials for modern technology, so it is not surprising that there is a rich literature about the formation of polymers. Most of that literature focuses on the formation of biopolymers and plastic but the polymerization models (Dotson *et al.*, 1996) for those cases are potentially useful to geochemists.

Many biopolymers and synthetic polymers are linear, which means that their constituent monomers are linked up in a head to tail fashion. For example, cellulose, $(C_6H_{10}O_5)_n$, is a linear array of glucose molecules, and polyethylene, $(C_2H_2)_nH_2$, is a linear grouping of ethylene molecules. However, polymers linked in three dimensions play a more important role in geochemical situations. For example, humic acids are randomly linked three-dimensional polymers consisting of various degradation products of biochemical substances.

Models for the formation of linear polymers are relatively simple (Dotson *et al.*, 1996). The chemical reaction for the growth of an $(AB)_n$ linear polymer involves hooking up additional $(AB)_m$ molecules head to tail.

Polymerization rates

$$(AB)_n + (AB)_m = (AB)_{n+m} \qquad (9.1)$$

The overall rate of this reaction depends upon the rate of interaction between the A and B functional groups with concentrations A and B.

$$\frac{dA}{dt} = \frac{dB}{dt} = -kAB \qquad (9.2)$$

Because the monomer AB contains equal amounts of A and B, $A = B$. At the initial conditions only the AB monomer exists. This species can be treated as a polymer (P) with a chain length of one so $A = B = P_o$. Equation (9.2) can be rewritten in terms of the concentration of polymer units (P).

$$\frac{dP}{dt} = -kP^2 \qquad (9.3)$$

Equation (9.3) can be integrated with the initial conditions of $P = P_o$ when $t = 0$ to find the concentration of P as a function of time.

$$P = \frac{P_o}{1 + ktP_o} \qquad (9.4)$$

The rate of decrease of monomer concentration (dP_1/dt) is proportional to the concentration of P_1 and P.

$$\frac{dP_1}{dt} = -2kP_1 P \qquad (9.5)$$

The factor of two in this equation arises because there are two distinguishable reactions, one for each end of the polymer. Additional equations can be written to find the rate of change in the concentration of polymers of each chain length (Dotson *et al.*, 1996). This model has didactic value, but most geochemical polymers are three-dimensional so they have more than two attachment points. Models of three-dimensional polymerization networks, such as occur with silica polymerization (Jin *et al.*, 2011), can become quite complex.

Example 9.1. Silica scaling

Deposition of amorphous silica (silica scale) onto working surfaces in geothermal power plants (Gunnarson and Arnórsson, 2003) and water desalinization plants (Weng, 1995) reduces rates of heat and mass transfer thereby reducing the efficiency of these enterprises. Silica scaling occurs when solutions become supersaturated with respect to amorphous silica because of cooling, evaporation of water, or a chemical reaction. The reaction between two monomers joins two Si atoms and eliminates a water molecule.

$$H_4SiO_4 + H_4SiO_4 \underset{}{\overset{K_1}{\rightleftharpoons}} H_6Si_2O_7 + H_2O \quad (9.6)$$

The next polymerization step follows the same pattern.

$$H_6Si_2O_7 + H_4SiO_4 \underset{}{\overset{K_2}{\rightleftharpoons}} H_8Si_3O_{10} + H_2O \quad (9.7)$$

The rate of monomer loss from solution (R, molal/sec) tends to be proportional to the fourth power of monomer concentration (Icopini *et al.*, 2005; Rothbaum and Rohde, 1979). Icopini *et al.* (2005) explain this by assuming that the rate-determining step is a reaction converting the trimer, formed by reaction (9.7), into a cyclical tetramer.

$$H_8Si_3O_{10} + H_4SiO_4 \xrightarrow{k_3} H_8Si_4O_{12} + 2H_2O \quad (9.8)$$

If reactions (9.6) and (9.7) reach equilibrium quickly, the overall rate is proportional to the concentration of $H_8Si_3O_{10}$, which is related to the monomer concentration by the K_1 and K_2 equilibrium constants. The $f(\gamma)$ term is the product of activity coefficients needed to convert the activity of the silica species into concentrations.

$$R = f(g)k_3 m_{H_8Si_3O_{10}} m_{H_4SiO_4} = f(g)k_3 K_1 K_2 \frac{m_{H_4SiO_4}^4}{a_{H_2O}^2} \quad (9.9)$$

Equation (9.9) can be simplified by representing the silica concentration as m and combining all the other terms into an apparent rate constant (k').

$$-\frac{dm}{dt} = k'm^4 \quad (9.10)$$

Equation (9.10) can be rearranged and integrated for the boundary condition that when $t = 0$, $m = m_o$.

$$\int_{m_o}^{m} \frac{dm}{m^4} = -k \int_0^t dt \quad (9.11)$$

$$-k't = \frac{1}{3}(m^{-3} - m_o^{-3}) \quad (9.12)$$

The concentration of monomer as a function of time is found by solving Eq. (9.12) for m.

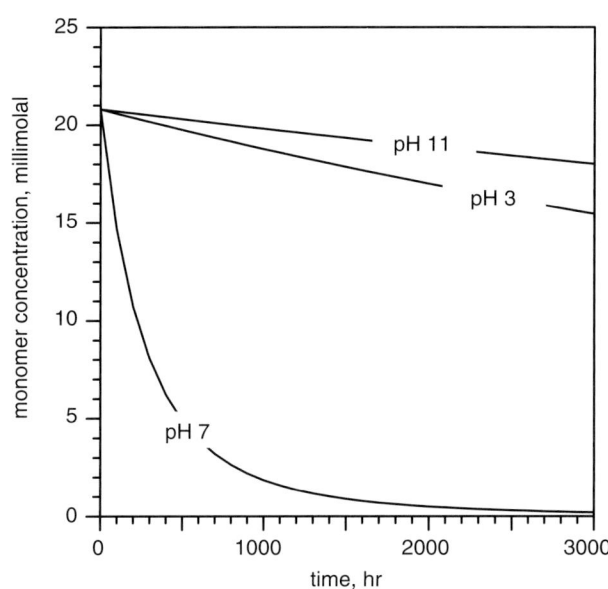

Figure 9.1. Silica monomer concentration as a function of time due to monomer incorporation into polymers. The formation of silica polymers is fast at near-neutral pH and slow at both high and low pH.

$$m = \left(\frac{1}{3k't + m_o^{-1/3}} \right)^3 \quad (9.13)$$

This equation, along with values of m_o and k' (Table 2 in Icopini et al., 2005), can be used to forward model the monomer concentration (m, millimolal) as a function of time (Figure 9.1) for an ionic strength of 0.01.

$m_o = 20.8$ mm (millimolal) $= 2.08 \times 10^{-2}$ m (same as for experiments in Icopini et al., 2005)

$k'(\text{pH} = 3) = 1.17 \times 10^{-9}$ mm^{-3}sec^{-1}
$k'(\text{pH} = 7) = 4.17 \times 10^{-8}$ mm^{-3}sec^{-1}
$k'(\text{pH} = 11) = 3.53 \times 10^{-10}$ mm^{-3}sec^{-1}

Nucleation rates

When a homogeneous aqueous phase becomes supersaturated with respect to a solid phase, the rate of appearance of the new solid phase is controlled by nucleation kinetics. Nucleation occurs whenever enough of the solute species come together to create a particle that can grow spontaneously. Classical models of nucleation kinetics are based on a combination of statistical mechanics and thermodynamic principles (Dunning, 1969; Ford, 2004; Leubner, 2000). All these models assume that the rate of appearance of nuclei is proportional to the equilibrium concentration of molecular clusters of a critical size. These models contrast with spinodal decomposition models that assume that fluctuations in the solution lead to the aggregation of solute molecules into a disordered cluster and the rate of formation of the solid depends upon the rate of rearrangement of molecules in the cluster to form a crystal (Gebauer *et al.*, 2008; Meldrum and Sear, 2008).

Solubility as a function of cluster size

Classical nucleation models are based on the concept that work is required to create the new interface bounding a growing molecular cluster. The amount of work (ΔG_c, J/cluster) needed to produce the cluster's interface is the product of the cluster's interfacial area (A, m^2) and the surface free energy of the interface (σ, mJ/m^2) (Adamson and Gast, 1997). This means that a cluster's surface free energy becomes more positive as it grows in size, but this increase is offset by a decrease in the cluster's bulk free energy of formation (ΔG_b, J/cluster), which is directly proportional to the number of molecules (n) in the cluster and the degree of supersaturation ($S = Q/K$). These two effects add together to give the overall free energy of formation of a cluster.

$$\Delta G_c = \Delta G_b + \Delta G_s = -nk_B T \ln S + \sigma A \tag{9.14}$$

This relationship can be recast to find a relationship between the solubility product of a cluster (K_c) with a diameter (D, m) and the solubility product of the bulk solid (K_b) (Berner, 1980).

$$\log K_c = \log K_b + \left(\frac{(2/3)b\sigma V_m}{2.303 RT}\right)\left(\frac{2}{D}\right) \tag{9.15}$$

The surface free energy of a cluster plays a critical role in determining its stability. Figure 9.2a shows that surface free energy of ionic solids correlates with the substance's bulk solubility (Söhnel, 1982) and Figure 9.2b shows how the solubility product of a cluster decreases with increasing cluster diameter until it approaches the solubility product of the bulk phase. A larger value of σ results in a higher solubility product for a particular cluster diameter. For clusters that are larger than about 1 μm the contribution of ΔG_s to the overall free energy is negligible, so their solubility is effectively the same as the bulk phase.

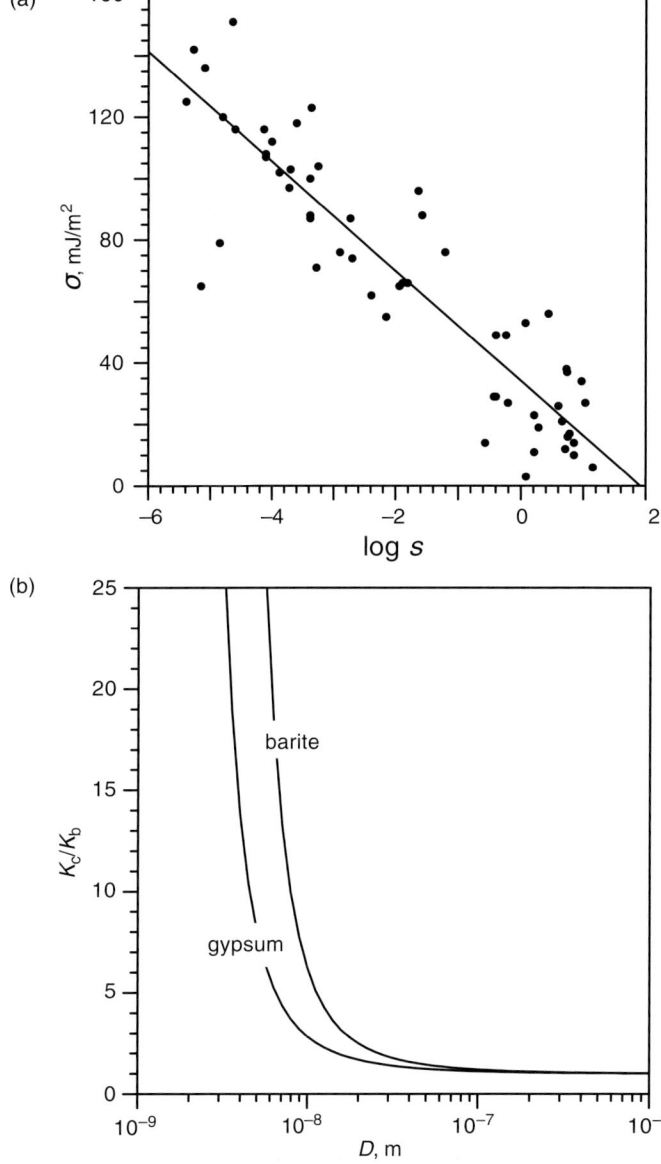

Figure 9.2. (a) Correlation between the interfacial free energy (σ, mJ/m^2) of ionic solids and the logarithm of their molar solubility (M_s, mol/L) at 25°C. The linear relationship is $\sigma = -17.8 \log M_s + 34.8$ (Söhnel, 1982). (b) Solubility product of gypsum (CaSO$_4$·2H$_2$O) (σ = 88 mJ/mol) and barite (BaSO$_4$) (σ = 136 mJ/mol) clusters relative to the solubility product of the bulk solid as a function of cluster diameter.

Nucleation

Nucleation models postulate that clusters must attain a critical size before they can irreversibly grow to a macroscopic size. Although details differ, all the classical nucleation models develop the same general relationship between the nucleation barrier, surface free energy, and degree of saturation. The models mostly differ in how the rate of addition of a monomer to the critical-sized cluster is derived from statistical mechanics (Dunning, 1955, 1969; Nielsen, 1964; Turnbull and Fisher, 1949; Uhlmann and Chalmers, 1966; Walton, 1969). The critical size barrier idea is cast in terms of the free energy of formation of the clusters making it analogous to the quasi-thermodynamic transition-state model presented in Chapter 5. Classical nucleation models all assume that the surface free energy is independent of cluster size. It is more likely that the surface free energy should be adjusted to account for its variation with cluster size (Fokin *et al.*, 2010; Larson and Garside, 1986).

With increasing solution concentration there is a tendency for dissolved species to interact with each other to form dimers, trimers, and larger clusters that are in dynamic equilibrium with the monomers. The concentration of each cluster size can be expressed by an equilibrium constant for each reaction step.

$$
\begin{aligned}
&A + A = A_2 \quad & K_2 = a_A^2/a_{A_2} \\
&A_2 + A = A_3 \quad & K_3 = a_A^3/a_{A_3} \\
&\vdots \\
&A_{n-1} + A = A_n \quad & K_n = a_A^n/a_{A_n}
\end{aligned}
\qquad (9.16)
$$

This derivation uses a molar (mol/L) concentration scale to maintain consistency with previous derivations. The model could be constructed using a molal (mol/kg) concentration scale to avoid complications that arise with the large changes in solution density associated with changing temperature or solute concentrations. The free energy of reaction for assembling n molecules into a cluster is found from the equilibrium constant for the reaction.

$$\Delta G_r^\circ = \Delta G_f^\circ(A_n) - n\Delta G_f^\circ(A) = -RT \ln K_n \qquad (9.17)$$

The free energy of formation of the clusters is related to the cluster concentration.

$$\Delta G_f^\circ(A_n) = -nRT \ln a_A - RT \ln K_n = -RT \ln M_{A_n} \qquad (9.18)$$

Dividing $\Delta G_r^\circ(A_n)$ by Avogadro's number (clusters/mol) gives the free energy of formation of a single cluster (ΔG_c, J/cluster). This gives a relationship between ΔG_c and the number of clusters per liter (c_{An}, clusters/L).

$$\Delta G_c = -k_B T \ln c_{A_n} \qquad (9.19)$$

The equilibrium cluster concentration must be known in order to evaluate the right-hand side of Eq. (9.19), but that presents a significant analytical difficulty because the cluster concentration is very low. This problem is overcome by realizing that the amount of work (ΔG_c) needed to accumulate n

molecules of A into a single cluster is the difference between the decrease in the bulk free energy (ΔG_b, J/cluster) due to adding more molecules to the cluster and the increase in the surface free energy (ΔG_s, J/cluster) due to the cluster's increasing surface area (Adamson and Gast, 1997). This means that ΔG_c can be expressed in terms of the number of molecules in each cluster (n_A) and the Boltzmann constant (k_B).

$$\Delta G_c = \Delta G_b + \Delta G_s = -nk_B T \ln S + A_c \sigma \tag{9.20}$$

S ($= Q/K$, no units) is the degree of saturation of the solution. The surface area of a cluster (A_c, m²/cluster) can be expressed in terms of the cluster's volume (V_c, m³/cluster) using the relationship: $A_c = (36\pi)^{1/3} V_c^{2/3}$. The number of molecules in the cluster is V_c divided by the volume of 1 molecule ($V_1 = V_m/N_A$, m³/molecule).

$$\Delta G_c = -nk_B T \ln S + (36\pi)^{1/3} n^{2/3} V_1^{2/3} \sigma \tag{9.21}$$

A graph of ΔG_c versus n displays a maximum (n^*), which corresponds to the number of molecules in a critical cluster that is in equilibrium with the supersaturated solution. The addition of one more molecule to this cluster will cause it to spontaneously grow to macroscopic size. The value of n^* is found from the derivative of Eq. (9.21).

$$\frac{d\Delta G_c}{dn} = -k_B T \ln S + (2/3)(36\pi)^{1/3} n_A^{-1/3} V_1^{2/3} \sigma \tag{9.22}$$

When $d\Delta G_c/dn = 0$, $n = n^*$ so Eq. (9.22) can be set equal to zero and solved for n^*.

$$k_B T \ln S = (2/3)(36\pi)^{1/3} (n^*)^{-1/3} V_1^{2/3} \sigma \tag{9.23}$$

$$n^* = \frac{(32\pi/3) V_1^2 \sigma^3}{(k_B T \ln S)^3} \tag{9.24}$$

The free energy of formation of the critical nucleus (ΔG_c^*, J/cluster) is found by substituting Eq. (9.24) into Eq. (9.21).

$$\Delta G_c^* = -\frac{(32\pi/3) V_1^2 \sigma^3}{(k_B T \ln S)^3} (k_B T \ln S) + (36\pi)^{1/3} \left(\frac{(32\pi/3) V_1^2 \sigma^3}{(k_B T \ln S)^3} \right)^{2/3} V_1^{2/3} \sigma \tag{9.25}$$

Equation (9.25) can be simplified using the following relationship.

$$(36\pi)^{1/3}(32\pi/3)^{2/3} = (6)^{2/3}\left(\frac{32}{3}\right)^{2/3}\pi = \left(\frac{192}{3}\right)^{2/3}\pi = (64)^{2/3}\pi = 16\pi = \frac{48\pi}{3} \quad (9.26)$$

This allows the right-hand side of Eq. (9.25) to be simplified to a single term.

$$\Delta G_c^* = -\frac{(32\pi/3)V_1^2\sigma^3}{(k_BT\ln S)^2} + \frac{(48\pi/3)V_1^2\sigma^3}{(k_BT\ln S)^2} = \frac{(16\pi/3)V_1^2\sigma^3}{(k_BT\ln S)^2} \quad (9.27)$$

The nucleation rate (N, nuclei/L sec) is the product of the flux of monomers (J_A, molecules/m² sec) to each critical nucleus and the surface area of each critical nucleus (A_c^*, m²/cluster) multiplied by the number of critical nuclei per unit volume of solution (c^*, nuclei/L).

$$N = J_A A_c^* c^* \quad (9.28)$$

Determining the flux of monomers to the critical nuclei (J_A, molecules/L sec) is a challenge. This flux has been modeled using various statistical mechanics approaches. These models estimate the flux of molecules from the solution to the surface of the critical nuclei but they include various adjustments such as the Zeldovich factor (Markov, 2003), which accounts for the possibility that some critical-sized clusters will dissolve to a smaller size instead of growing into crystallites. Presently there is no widely accepted model for J_A, so Eq. (9.28) is usually recast into a semi-empirical form, which combines the monomer flux and cluster surface area into a pre-exponential term (Ω, crystallites/L sec).

$$N = \Omega c^* \quad (9.29)$$

Rearranging Eq. (9.19) gives the concentration of critical-sized clusters.

$$c^* = e^{-\Delta G_c^*/k_BT} \quad (9.30)$$

Combining Eqs (9.27), (9.29), and (9.30) produces an equation that expresses the nucleation rate in terms of the degree of saturation.

$$N = \Omega e^{(-\Delta G_c^*/k_BT)} = \Omega e^{\left(-\frac{(16\pi/3)V_1^2\sigma^3}{(k_BT\ln S)^2}/k_BT\right)} = \Omega e^{\left(-\frac{(16\pi/3)V_1^2\sigma^3}{k_BT(k_BT\ln S)^2}\right)} \quad (9.31)$$

This equation can be log-transformed to give log N in terms of log S.

$$\log N = \log\Omega - \frac{(16\pi/3)V_1^2\sigma^3}{(\ln 10 k_BT)^3}(\log S)^{-2} \quad (9.32)$$

Experimental results suggest that log Ω is on the order of 33 ± 3 for nucleation in aqueous solution at 25°C (Nielsen, 1964; Pina and Putnis, 2002).

During their initial growth stages, the clusters are likely to have structures that are more disordered than that of the bulk solid (Garten and Head, 1970). This idea has been incorporated into models that posit the formation of a disordered cluster as an initial step followed by the formation of a crystallite in a second step (Erdemir et al., 2009). These models are more complicated conceptually and quantitatively than the classical model presented here. Alternatively, Söhnel and Garside (1988) deal with this possibility by making the surface free energy of the clusters a function of their size.

Example 9.2. Barite (BaSO$_4$) nucleation rate

Barite scales can clog water treatment, petroleum production, and geothermal facilities (Boerlage et al., 2002; He et al., 1994). An additional problem is that the barite that precipitates in these and other systems often incorporates significant amounts of radioactive elements, especially radium. Removal and disposal of these radioactive barite precipitates is complicated and costly (Ceccarello et al., 2004). Schemes to mitigate barite scale formation typically use additives to inhibit barite nucleation. The first step toward evaluating the efficacy of these additives is to develop a model for barite nucleation rates for untreated conditions. Most of these additives are chosen because they increase the surface free energy of the barite clusters thus slowing the nucleation rate.

The following data are needed to construct a model of barite nucleation.

$\sigma = 136$ mJ/m^2 = 0.136 J/m^2
$V_m = 5.21 \times 10^{-5}$ m^3/mol so $V_1 = 8.65 \times 10^{-29}$ m^3/molecule
$k_B = 1.38 \times 10^{-23}$ J/molecule K
$(36\pi)^{1/3} = 4.84$
$32\pi/3 = 33.5$
$16\pi/3 = 16.8$
$T = 298$ K

The first step toward developing a conceptual model of the nucleation process uses Eq. (9.21) to map out ΔG_c as a function of log S. It is convenient to recast this equation by inserting the appropriate numerical constants.

$$\Delta G_c = -n(1.38 \times 10^{-23})(298) 2.303 \log S + 4.84 n^{2/3} (8.65 \times 10^{-29})^{2/3} (136 \times 10^{-3})$$
$$= -n(9.47 \times 10^{-21}) \log S + n^{2/3}(1.29 \times 10^{-19})$$

(9.33)

The maximum on each graph of ΔG_c versus n (Figure 9.3a) occurs where $n = n^*$. These graphs show that the ΔG_c^* values associated with each maximum decrease

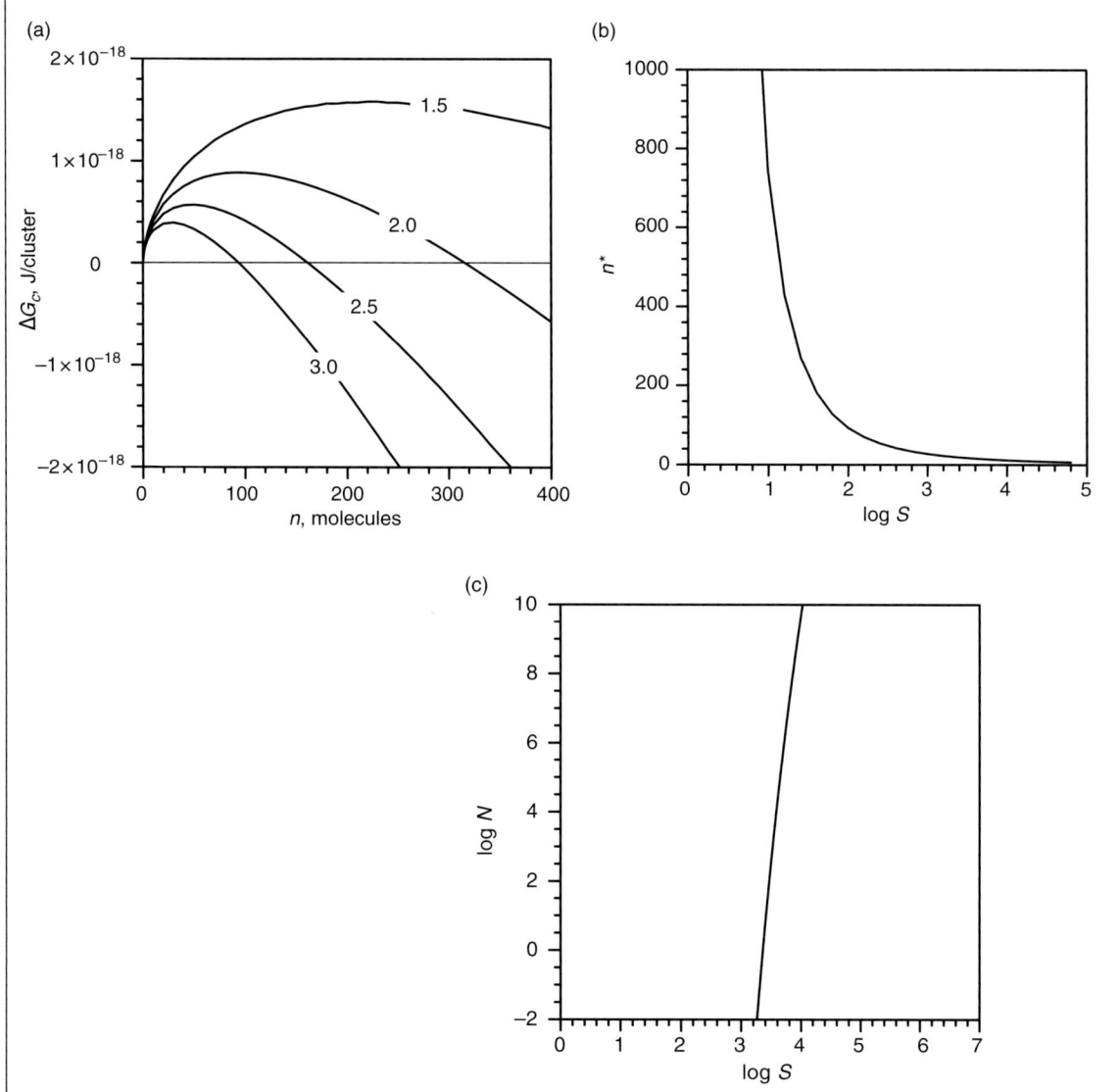

Figure 9.3. (a) Free energy of formation of $BaSO_4$ clusters as a function of the number of $BaSO_4$ molecules in the cluster contoured for values of log S ranging from 1.5 to 3.0. (b) Number of $BaSO_4$ molecules in the critical cluster as a function of log S. (c) Nucleation rate (nuclei/L sec) of $BaSO_4$ as a function of log S.

as log S increases, which means that the barrier to nucleation decreases as the solution becomes more supersaturated.

The size of the critical nucleus as a function of the degree of saturation is found using Eq. (9.34).

$$n^* = \frac{(33.5)(8.65 \times 10^{-29})^2 (0.136)^3}{(1.38 \times 10^{-23})^3 (298)^3 (2.303)^3 (\log S)^3} = \frac{742.7}{(\log S)^3} \quad (9.34)$$

This equation shows that the size of the critical nucleus grows very large when the degree of saturation drops below 10 ($\log S < 1$) and becomes very small when the degree of saturation exceeds 10,000 ($\log S > 4$) (Figure 9.3b).

The barite nucleation rate as a function of the degree of saturation is calculated using Eq. (9.32).

$$\log N = 33 - \frac{(16.8)(8.65 \times 10^{-29})^2 (0.136)^3}{((2.303)(1.38 \times 10^{-23})(298))^3} (\log S)^{-2} = 33 - 373(\log S)^{-2} \quad (9.35)$$

Figure 9.3c shows that the log N versus log S curve is very steep, so that homogeneous nucleation rates are very low when log S is less than 3 but they become very fast as log S increases from 3 to 4. When log S is less than about 3, barite nucleates so slowly that it is unlikely to cause a technological problem. Barite formation for log $S < 3$ will involve heterogeneous nucleation (Kashchiev and van Rosmalen, 2003), which is modeled using a modified geometry and surface free energy for the critical cluster.

Aggregation rates

If the concentration of polymers or crystallites is high enough they can aggregate into larger clusters. For the simplest case, each cluster (A) moves around in the solution as the result of Brownian motion (diffusive behavior) and whenever two clusters come into contact they stick together. This is called diffusion-limited aggregation (DLA). For clusters of equal diameter, the concentration of A decreases following a second-order rate equation.

$$\frac{dc_A}{dt} = -\frac{k_S}{2} c_A^2 \quad (9.36)$$

The rate constant for this equation can be calculated from the Smoluchowski model.

$$k_S = \frac{8 k_B T}{3 \eta} \quad (9.37)$$

The Boltzmann constant (k_B) equals 1.38×10^{-23} J/molecule K and at 25°C the dynamic viscosity of water (η) equals 8.91×10^4 Pa sec (1 Pa = 1 J/m³).

$$k_S = \frac{8k_BT}{3\eta} = \frac{8(1.38 \times 10^{-23})(298)}{3(8.91 \times 10^{-4})} = 1.23 \times 10^{-17} \frac{m^3}{\text{cluster sec}} \quad (9.38)$$

This constant can be inserted into Eq. (9.36) to predict the rate of disappearance of A due to diffusion-limited aggregation at 25°C.

$$\frac{dc_A}{dt}\left(\frac{\text{clusters}}{m^3 \text{ sec}}\right) = -\left(\frac{1}{2}\right)k_S c_A^2 = -\left(\frac{1}{2}\right)1.23 \times 10^{-17} c_A^2 \quad (9.39)$$

Equation (9.36) can be integrated from the initial cluster concentration ($c_A{}^\circ$) to the time when $c_A = \frac{1}{2}c_A{}^\circ$.

$$-\frac{2}{c_A} + \frac{2}{c_A^o} = -\frac{2}{1/2 c_A} + \frac{2}{c_A^o} = -k_S t_{1/2} \quad (9.40)$$

Equation (9.40) can be rearranged to find the time needed to reduce the cluster concentration to half its original value.

$$t_{1/2} = \frac{2}{k_S c_A^o} \quad (9.41)$$

Equation (9.41) shows that cluster aggregation is very sensitive to the cluster concentration, so that for low concentrations very long times are required for significant aggregation to occur (Figure 9.4a) but aggregation times are very short for suspensions with high cluster concentrations. Because diffusion rates increase with increasing temperature, the aggregation times decrease significantly with increasing temperature (Figure 9.4b). When log k_S is graphed versus $1/T$ (Figure 9.4c), the activation energy for aggregation is found to be ~15.5 kJ/mol, which is consistent with a diffusion-limited process.

Diffusion-limited aggregation (DLA) is often referred to as a fast process because the above model predicts the maximum aggregation rate. Reaction-limited aggregation (RLA) rates are slower than predicted by the DLA model. The effect of the slow attachment reaction rate on the aggregation rate is expressed as a stability ratio (W) and is defined as the ratio of the Smoluchowski rate constant (k_S) to the observed rate constant (k_O).

$$W = \frac{k_S}{k_O} \quad (9.42)$$

Figure 9.5 shows that W declines dramatically with increasing ionic strength because the ions cause the double layer to collapse, which decreases the electrostatic repulsion between the colloid particles.

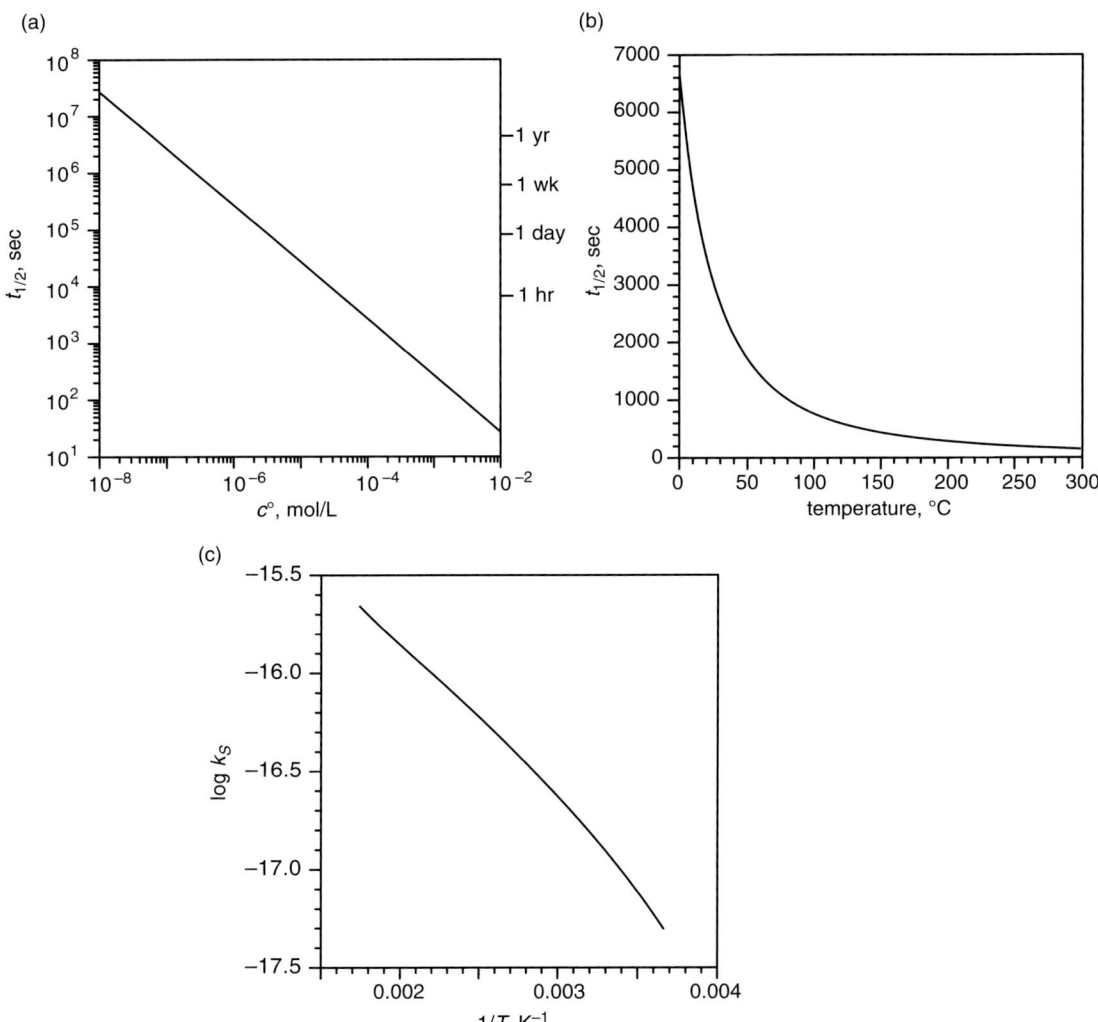

Figure 9.4. (a) DLA model prediction of the time needed for half the clusters to aggregate as a function of cluster concentration. (b) Time needed for half the clusters to aggregate as a function of temperature for $c° = 1 \times 10^{-4}$ mol/L. (c) Arrhenius plot showing log k_S as a function of $1/T$ for a DLA process.

Example 9.3. Aggregation time for silica clusters

Axford (1997) showed that the W value for silica cluster aggregation is affected by the ionic strength and cluster size. This information can be used to model the effect of ionic strength and cluster size on the aggregation time.

The first step in building this model uses multiple linear regression to fit the W data in Figure 9.5 as a function of ionic strength (I) and cluster size (N).

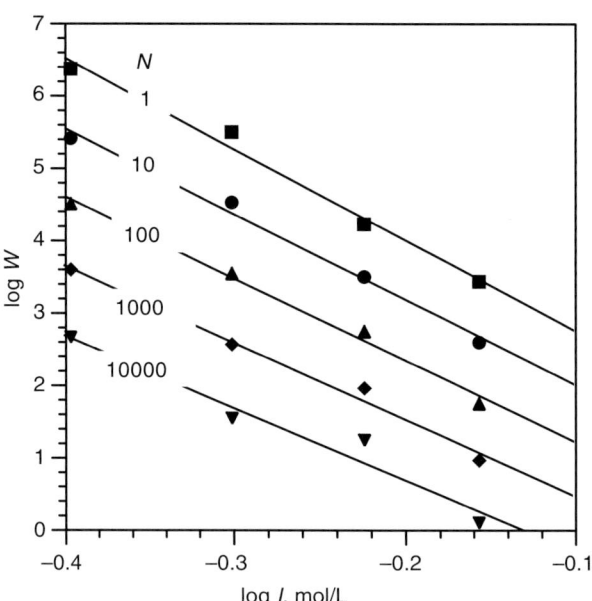

Figure 9.5. Logarithm of experimentally determined stability ratios for colloidal silica ($c_A^\circ = 6 \times 10^{16}$ clusters/m^3) as a function of the logarithm of the ionic strength (I = molarity of KCl) contoured in aggregate size expressed as the number (N) of 12 nm diameter unit clusters in the aggregate (pH = 9.4) (Axford, 1997).

$$\log W = 1.85 - 11.2 \log I - 0.872 \log N \tag{9.43}$$

The next step substitutes $\log k_S$ into the logarithmic transform of Eq. (9.42) to find an empirical rate constant (k_o).

$$\log k_o = \log k_S - \log W = -16.9 - \log W \tag{9.44}$$

Then Eq. (9.41) is recast in terms of the empirical rate constant (k_o).

$$\log t_{1/2} = 0.301 - \log c_A^\circ - \log k_o \tag{9.45}$$

Combining Eqs (9.43), (9.44), and (9.45) gives a function for $t_{1/2}$ as a function of cluster size, ionic strength, and initial cluster concentration.

$$\log t_{1/2} = 19.1 - \log c_A^\circ - 11.2 \log I - 0.872 \log N \tag{9.46}$$

Figure 9.6 shows that $t_{1/2}$ is very small for large clusters at high ionic strength and very large for small clusters at low ionic strength. The silica cluster aggregation rate is also strongly affected by pH, which controls the surface charge of the clusters.

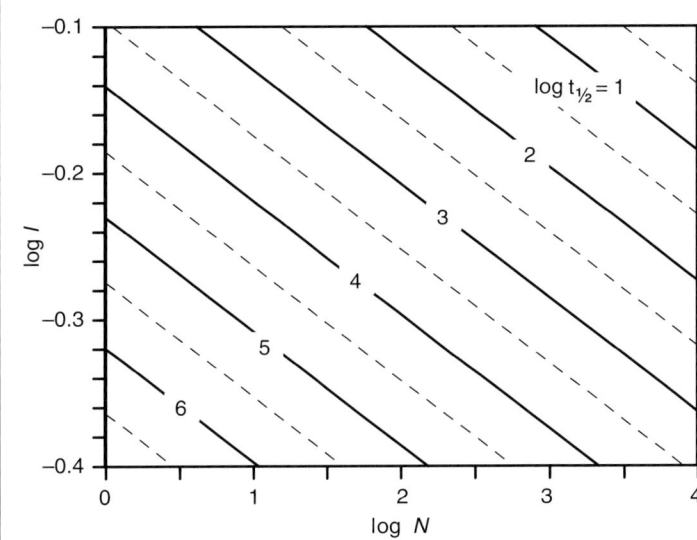

Figure 9.6. Logarithm of the time needed for one half of the silica clusters to aggregate (log $t_{1/2}$, time in seconds) contoured in terms of the logarithm of the ionic strength (I) and logarithm of the cluster size (N = number of 12 nm diameter units in the cluster).

The previous discussion assumed perikinetic conditions where the contact rate of the clusters is limited to collisions due to their Brownian motion. Orthokinetic aggregation refers to situations where fluid motion in the suspension due to stirring or fluid flow increases the contact rate, see for example Edzwald et al. (1974). There are many good books that cover this, and related topics, in more detail, e.g. Meyer (1999). Burd and Jackson (2009) review the principles that apply to aggregation in marine settings. Further complexity in the aggregation process arises when the clusters interact via asymmetric force fields (Bishop et al., 2009), which causes oriented aggregation. Oriented aggregation appears to be a common geochemical process that plays a role in the formation of many minerals (Burrows et al., 2010; Penn, 2004; Penn et al., 2007).

Solid–solid transformation rates

The rate of transformation of a metastable solid (parent) phase (A) to form a more stable solid (product) phase (B) is usually modeled using the Avrami equation (Avrami, 1939, 1940), which is also known as the Johnson–Mehl–Avrami–Kolmogorov (JMAK) equation. This equation is based on a model that assumes that the transformation involves the nucleation of the product phase followed by its growth until the parent phase is replaced by the product phase. This model is based on three assumptions: (1) nucleation occurs homogeneously throughout the parent phase, (2) the growth rate is constant over the entire time of the transformation, and (3) the growth rate of the new phase is the same in all directions.

The JMAK equation predicts the fraction of parent phase transformed ($\alpha = V/V_t$, m³/m³), which is the volume of parent transformed divided by the total volume, as a function of time. In the initial stages of the transformation, the product phase nucleates and grows at random places throughout the parent phase. The product phase has a spherical shape because it grows at the same rate in all directions. Eventually the growing spheres overlap and there is no nucleation or growth in the overlapped volume. If this overlap did not occur, the extended volume of transformed material would be V_e and the fraction of material transformed would be α_e ($= V_e/V$, m³/m³). Once the product phases overlap, the rate of change of α must be adjusted for the overlap of the product phase regions.

$$d\alpha = d\alpha_e (1-\alpha) \tag{9.47}$$

For the case where abundant nuclei are present in the initial stages of the transformation, which is termed site saturation, the extended volume of one sphere of product phase is modeled by changing the sign for the rate constant in the shrinking particle model (Chapter 6) to make it a growing particle model. The linear growth rate of each sphere of the product phase is G (m/sec) and its volume is V (m³).

$$V_i = \frac{4\pi}{3}(Gt)^3 \tag{9.48}$$

The extended fraction of transformed material is found by multiplying Eq. (9.48) by the number of nuclei in the total volume N ($=$ nuclei/V_t).

$$\alpha_e = \frac{\sum V_i}{V_t} = N\frac{4\pi}{3}(Gt)^3 \tag{9.49}$$

The rate of change of the extended fraction of transformed material is found by taking the time derivative of Eq. (9.49).

$$d\alpha_e = N4\pi(Gt)^2 dt \tag{9.50}$$

Equation (9.50) is substituted into Eq. (9.47) to eliminate the α_e term.

$$d\alpha = (1-\alpha)N4\pi(Gt)^2 dt \tag{9.51}$$

Equation (9.51) can be rearranged and integrated from $t = 0$ to t.

$$\int_0^\alpha \frac{d\alpha}{(1-\alpha)} = N4\pi G^2 \int_0^t t^2 dt \qquad (9.52)$$

$$-\ln(1-\alpha) = \frac{N4\pi G^2}{3} t^3 = kt^3 \qquad (9.53)$$

Equation (9.53) is transformed and rearranged to find α as a function of t.

$$\alpha = 1 - e^{-kt^3} \qquad (9.54)$$

A similar derivation, which assumes that the transformation rate is controlled by the nucleation rate, gives an equation where t is raised to the fourth power. The classical Avrami equation generalizes these results by assuming that α is related to t by an equation with an arbitrary value for n.

$$\alpha = 1 - e^{-kt^n} \qquad (9.55)$$

This equation is transformed to a linear form to extract values of k and n from experimental data.

$$\ln(-\ln(1-\alpha)) = \ln k + n \ln t \qquad (9.56)$$

Values of n found from experimental results are sometimes interpreted to identify the predominant reaction process. For example, if $n \approx 4$ the transformation rate is said to be limited by nucleation rates and if $n \approx 3$ the transformation rate is said to be limited by the growth rate of the product phase.

A modified version of the Avrami raises both k and t to the power of n, which gives k more conventional units of $\sec^{-1}$ rather than the units of $\sec^{-n}$ that are required for the classical Avrami equation (Maffezzoli et al., 1995).

$$\alpha = 1 - e^{-(kt)^n} \qquad (9.57)$$

The linearized version of the modified Avrami model is slightly different from Eq. (9.56).

$$\ln(-\ln(1-\alpha)) = n \ln k + n \ln t \qquad (9.58)$$

Although criticized (Marangoni, 1998), this modified version is widely used and it is important to be aware that the values of n and k are not interchangeable between the two forms (see Figure 9.7). Both forms predict "s"-shaped curves for graphs of α versus t. There is no theoretical justification for interpreting reaction processes from values of n found using the modified Avrami equation.

It is often useful to express the conversion in terms of a characteristic time. The classical Avrami equation (9.55) can be solved to express t as a function of α.

$$t = \left(-\frac{\ln(1-\alpha)}{k}\right)^{1/n} \tag{9.59}$$

Substituting 0.5 for α in Eq. (9.59) gives the time to 50% conversion.

$$t_{1/2} = \left(-\frac{\ln(1-0.5)}{k}\right)^{1/n} = \left(\frac{0.693}{k}\right)^{1/n} \tag{9.60}$$

The modified Avrami equation (9.57) can be solved to express t as a function of α.

$$t = \frac{1}{k}\left(-\ln(1-\alpha)\right)^{1/n} \tag{9.61}$$

Substituting 0.5 for α in Eq. (9.61) gives the time to 50% conversion.

$$t = \frac{1}{k}\left(-\ln(1-\alpha)\right)^{1/n} = \frac{1}{k}(0.693)^{1/n} \tag{9.62}$$

TTT diagrams

Time–temperature–transformation (TTT) diagrams are extensively used in materials sciences to map out the characteristic time for solid–solid transformations. They are sometimes called kinetic phase diagrams. These diagrams show the extent of transformation, α, contours on graphs with log t on the x-axis and either T or $1/T$ on the y-axis. They are often prepared by direct plotting of experimental data but they can be constructed for any type of reaction using the integrated rate equation along with a temperature function (Arrhenius equation) for the rate constant.

The simplest case is a TTT diagram based on a zeroth-order rate equation.

$$\frac{dn}{dt} = -k \tag{9.63}$$

The integrated form of this equation gives the number of moles remaining at a given time.

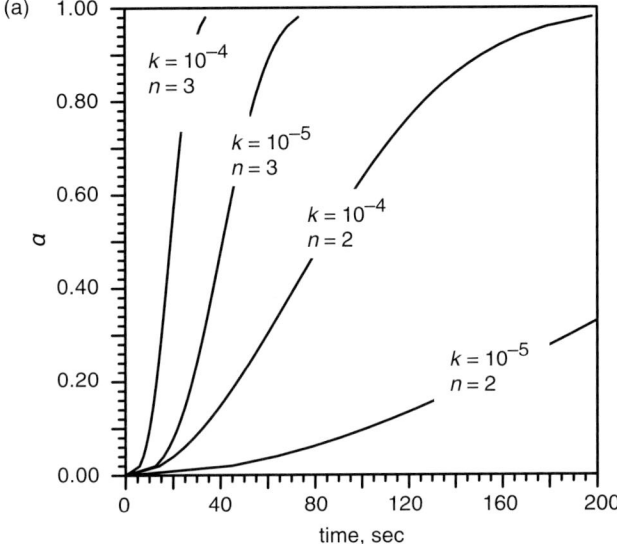
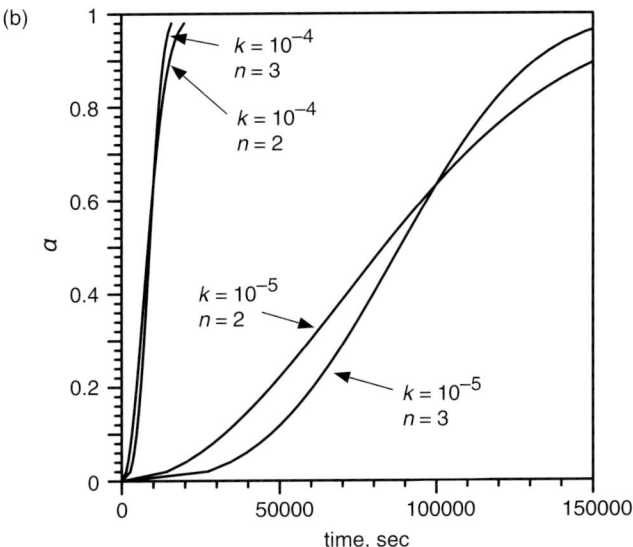

Figure 9.7. (a) Examples of the effect of changing k and n on the extent of the transformation versus time pattern predicted by the classical Avrami equation. Increasing k or decreasing n decreases the slope of the curve. (b) Examples of the effect of changing k and n on the extent of the transformation versus time pattern predicted by the modified Avrami equation. Increasing k decreases the slope of the curve but changing n has very little effect on the slope. Comparison of the $t_{1/2}$ values for $k = 1 \times 10^{-5}$ and $n = 3$ between models illustrates their significant quantitative difference.

$$n - n_o = -kt \qquad (9.64)$$

Equation (9.64) can be expressed in terms of the fraction remaining (p), or better yet in terms of the fraction transformed (α).

$$1 - \frac{n}{n_o} = 1 - p = \alpha = \left(\frac{k}{n_o}\right)t \qquad (9.65)$$

Equation (9.65) is transformed to give ln α as a function of ln t.

$$\ln \alpha = \ln k - \ln n_o + \ln t \qquad (9.66)$$

The log-transformed Arrhenius equation is substituted into Eq. (9.66) to get a function that expresses α as a function of T and t.

$$\ln \alpha = \ln A - \left(\frac{E_a}{R}\right)\frac{1}{T} - \ln n_o + \ln t$$

$$= \ln\left(\frac{A}{n_o}\right) - \left(\frac{E_a}{R}\right)\frac{1}{T} + \ln t \qquad (9.67)$$

$$\log \alpha = \log\left(\frac{A}{n_o}\right) - \left(\frac{E_a}{2.303R}\right)\frac{1}{T} + \log t$$

Although the TTT diagrams typically show T as a function of log t, it is more convenient to rearrange this equation to solve for log t as a function of $1/T$.

$$\log t = -\log\left(\frac{A}{n_o}\right) + \log \alpha + \left(\frac{E_a}{2.303R}\right)\frac{1}{T} \qquad (9.68)$$

In this case, the fraction-transformed boundary will be a straight line on a graph of $1/T$ versus log t. A graph of T versus log t can be constructed by taking the reciprocal of Eq. (9.68). That graph will display a curved line.

The TTT equation for the classical Avrami model is developed from Eq. (9.56).

$$\ln(-\ln(1-\alpha)) = \ln A - \left(\frac{E_a}{R}\right)\frac{1}{T} + n \ln t \qquad (9.69)$$

Equation (9.69) is rearranged so log t is a function of $1/T$.

$$\ln t = \left(\frac{1}{n}\right)\left(-\ln A + \ln(-\ln(1-\alpha)) + \left(\frac{E_a}{R}\right)\frac{1}{T}\right)$$

$$\log t = \left(\frac{1}{n}\right)\left(-\log A + \log(-2.303\log(1-\alpha)) + \left(\frac{E_a}{2.303R}\right)\frac{1}{T}\right) \qquad (9.70)$$

The TTT equation for the modified Avrami model (Eq. (9.58)) is slightly different.

$$(1/n)\ln(-\ln(1-\alpha)) = \ln A - \left(\frac{E_a}{R}\right)\frac{1}{T} + \ln t \qquad (9.71)$$

Equation (9.71) is rearranged so log t is a function of $1/T$.

$$\ln t = -\ln A + (1/n)\ln\left(-\ln(1-\alpha)\right) + \left(\frac{E_a}{R}\right)\frac{1}{T}$$
$$\log t = -\log A + (1/n)\log\left(-2.303\log(1-\alpha)\right) + \left(\frac{E_a}{2.303R}\right)\frac{1}{T} \quad (9.72)$$

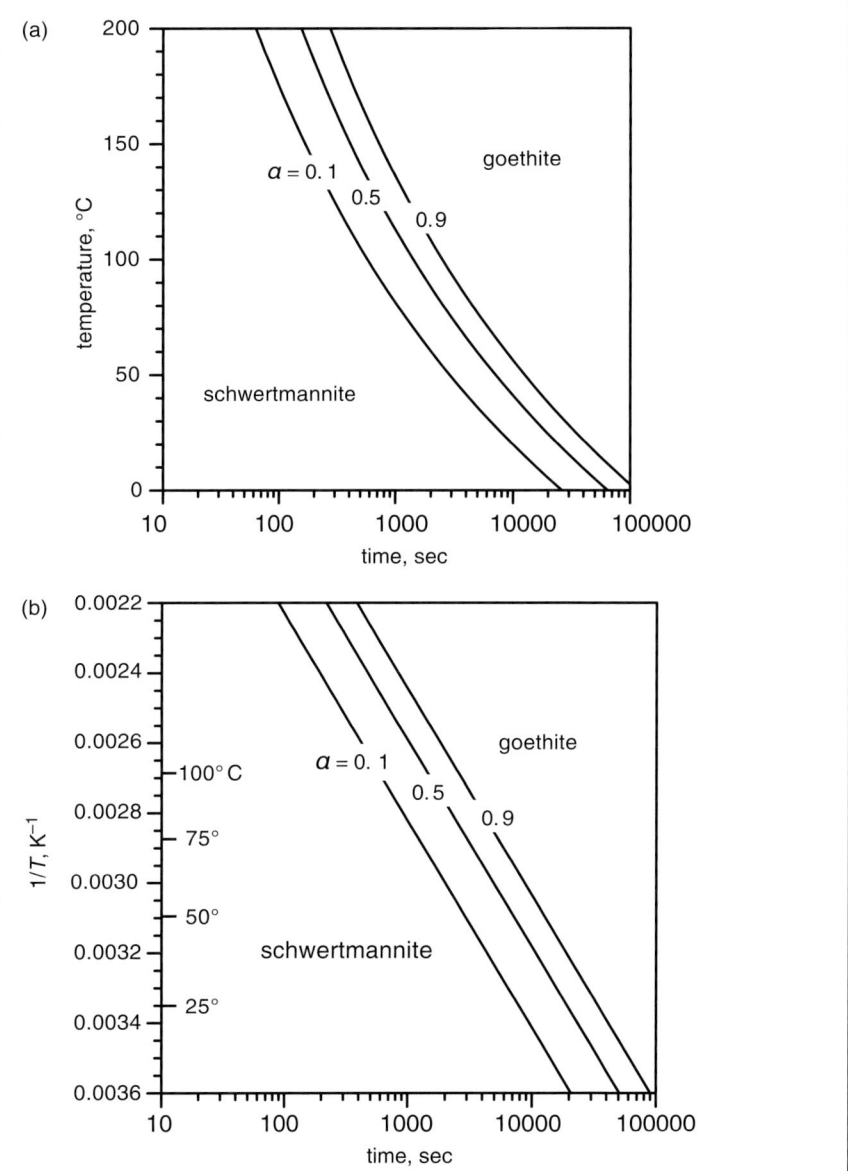

Figure 9.8. TTT diagrams contoured in the extent of transformation of schwertmannite to goethite based on data from Davidson *et al.* (2008). (a) When a linear temperature axis is used the contours are curved lines. (b) When a reciprocal temperature axis is used the contours are straight lines.

Example 9.4. Schwertmannite to goethite transformation

Schwertmannite (*sch*) is a fairly common mineral in acid mine drainage settings. It is metastable and eventually transforms into goethite (*goe*).

$Fe_8O_8(OH)_{8-2x}(SO_4)_x$ (*sch*) + $2xH_2O$ = $8FeOOH$(*goe*) + $2xH^+$ + xSO_4^{2-}

Davidson *et al.* (2008) measured the rates of this transformation over the temperature range of 60 to 190°C and fit the results to the *modified* Avrami equation. The average value of n from their fits was 2.1(0.25). Fitting their reported rate constants to the Arrhenius equation gives an estimated value of E_a of 32.4 kJ/mol with log A = 1.31. These values are substituted into Eq. (9.72) to create the TTT diagram shown in Figure 9.8.

Chapter 10
Pattern formation

So far this book has offered appetizers. This chapter deals with the preparation of the main course by introducing concepts for linking the simple models discussed in this book to understand the complex processes that lead to pattern formation in geological settings. Geoscientists spend much of their time and effort identifying and explaining naturally occurring spatial and temporal patterns with the goal of interpreting those patterns to understand the processes and conditions that formed them. They are challenged by the need to identify meaningful patterns in situations that often appear to be chaotic (Crutchfield, 2012). Meaningful patterns are frequently subtle and recognizing them often requires clues provided by process models. Most pattern-forming systems are too complex to interpret in a holistic way so our strategy is to first parse them into simple processes, which can be accurately modeled using methods like those explained in this book. The resulting models of discrete processes are then linked to simulate the overall pattern-forming scenario. This strategy is widely used in science and technology. For example, engineers design processing plants by dividing the overall process into unit operations, each of which is responsible for a single chemical or physical transformation of a feedstock (Gupta and Yan, 2006; Hendricks, 2006; McCabe et al., 1993). These unit operations are modeled separately and the models are combined to simulate the entire processing plant. This strategy is especially effective when the unit operations occur in a linear array of steps so that the product of one step is the feed for the next. Natural processes are often more complicated because they can switch from one path to another in a stochastic manner or because there is one or more feedback loops in the overall process. Regardless of the complexity of the situation, the divide and analyze strategy is the most effective way to understand how observed patterns are related to unit processes. The challenge of interpreting pattern-forming processes is a very exciting scientific frontier (Ball, 1999; Nicolis and Prigogine, 1989).

Geochemists observe patterns of element and mineral distribution in nature and then use forward and inverse models of unit processes to understand how these patterns developed. The patterns result from the interplay between energy flow and feedback in a network of unit processes, which means that the element and mineral distribution patterns preserve valuable information about their conditions of formation. The reductionist strategy of dividing the overall process into simple units is the first step in interpreting the overall process. The second step is recognizing and understanding emergent behaviors caused by the feedback loops. Non-equilibrium thermodynamics and the Turing model offer powerful insights into the behavior of linked energy and mass flow processes with feedback.

Non-equilibrium thermodynamics

Non-equilibrium thermodynamics (de Groot and Mazur, 1984; Prigogine, 1977) affords an abstract, and therefore very general, foundation for understanding pattern formation. The abstract nature of this approach makes direct application to geochemical problems difficult, but non-equilibrium thermodynamics does provide a powerful conceptual basis for thinking about pattern formation.

Non-equilibrium thermodynamics uses the notion of rate of entropy production (Kondepudi and Prigogine, 1998) to show that physical and chemical transformations can all be described in terms of forces and fluxes. For example, the force driving a chemical reaction is the difference between the chemical potential of the reactants and products. The greater this difference, the faster the flux (rate) of conversion of reactants into products. The force driving diffusion is the difference between a species' chemical potentials in two spatial regions. The diffusion flux from the region of high chemical potential to the region of low chemical potential is directly proportional to this force. Similar relationships exist between the rate of flow of electrical current and voltage differences and the rate of heat flow and temperature differences. In all these cases, a resistance limits the rate of transformation or mass transfer. This resistance shows up in rate equations as the rate constant and in diffusion equations as the diffusion coefficient divided by distance. This means that all these entropy-producing processes are analogous at a very fundamental level.

An even more potent concept comes from the Onsager reciprocal relations, which states that there is a coupling between conjugate force–flux pairs. For example, mass transfer of species by diffusion in an aqueous solution causes a change in concentration, which is accompanied by heat consumption or release due to the heat of dilution. This sets up a thermal gradient, which causes heat flow. The resulting link between heat flux and isothermal diffusion is the Dufour effect. Its conjugate is the Soret effect, which is the diffusional mass flux linked to heat flow. The Soret effect has been coupled with gravitational effects to separate chemical species and isotopes in chemical technology (Platten *et al.*, 2003; Powers and Wilke, 1957). It may be important in certain geotechnical applications (Vidal and Murphy, 1999) and it appears to cause separation of chemical species and isotopes in some geological settings (Cygan and Carrigan, 1992).

The relationship between forces and fluxes is simple only when the driving force is small. As the driving force grows, this relationship becomes increasingly nonlinear. This effect, combined with the coupling of processes, leads to some surprising behaviors. The most important of these is spontaneous self-organization (Nicolis and Prigogine, 1977; Prigogine, 1980; Prigogine and Stengers, 1984). This means that many of the patterns observed in geological settings are the result of feedback in complex networks of nonlinearly interacting forces and fluxes. Even though we can usually model the unit processes involved, linking these models to predict the pattern-forming behavior is challenging. Nonetheless, non-equilibrium thermodynamics provides a quantitative and conceptual foundation for understanding pattern formation.

Turing model

The coupling of individual processes, either due to Onsager reciprocal type relationships or due to continuity requirements, produces feedbacks between unit processes. Feedback effects control the location and structure of many patterns. Feedback occurs when the output of a succeeding unit process somehow affects the behavior of an antecedent unit process. Turing (1953) quantified the relationship between feedback and pattern formation in biological systems and the Turing model has proved useful for explaining many kinds of biological patterns (Kondo and Miura, 2010). It has great potential for explaining many other kinds of pattern formation such as oscillating chemical reactions and Liesegang bands. The Briggs–Rauscher reaction, the Bray–Liebhafsky reaction, and the Belousov–Zhabotinsky reaction are well-known examples of a growing number of recognized oscillating chemical reactions (Epstein *et al.*, 1983; Field and Schneider, 1989) that exhibit cyclical temporal patterns. These same reactions can also produce combined temporal and spatial patterns, which appear as waves that move through the reaction medium (Ross *et al.*, 1988; Winfree, 1972). Liesegang (1913) recognized the occurrence of spatial patterns caused by precipitation linked to diffusion and used this idea to explain band formation in agates (Liesegang, 1915). Although agate bands are probably not a good example of this phenomenon (Heaney, 1993), Liesegang bands do occur in many geological settings (Augustithis *et al.*, 1980; Fu *et al.*, 1994) and they are relatively easy to produce in laboratory settings (Henisch, 1988; Krug *et al.*, 1996; Müller *et al.*, 1982). Feedback processes that link mass and energy fluxes are a key feature of pattern formation.

Geological examples

Porphyry copper ore deposits (John, 2010) are a good example of how energy and mass flux linked by feedback produce recognizable geologic patterns. Heat from the mantle produces copper-enriched magma by partial melting in the upper mantle or at the base of the crust. Because the magma has a lower density than the surrounding rocks, it rises carrying along this thermal energy until it is emplaced near the Earth's surface where cooling and crystallization release heat to the surroundings. The crystallizing magma also releases copper-rich hydrothermal fluids with sufficient pressure to hydrofracture a well-defined region near the crystallizing magma. The shape of this region is controlled by a dynamic interaction between the magmatic fluids and the surrounding meteoric water (Weiss *et al.*, 2012). The cooling hydrothermal solutions permeate the fractures where changes in pressure and temperature cause the precipitation of copper-bearing minerals in enough abundance to form an economic deposit. The porphyry copper process is driven by energy transport from the upper mantle to the Earth's surface and this energy flow does work, some of which concentrates copper. The location of the copper-enriched zone is fixed by feedback between the fluid flow, permeability generation, and chemical precipitation. These linked processes produce a distinct pattern of igneous features, rock alteration, isotopic distribution, etc., which form the basis for the exploration models used to discover new deposits (John, 2010). Geothermal energy drives this aspect of the porphyry copper pattern.

A subsequent phase of porphyry copper deposit pattern formation is driven by solar energy, via the hydrologic cycle. The hydrologic cycle transports water from the oceans to the location of a porphyry copper deposit. The resulting erosion exposes the sulfide-rich porphyry copper ore to the atmosphere allowing the copper sulfides to oxidize and release copper into solution. Some of the soluble copper is dispersed by flowing surface and groundwater to produce copper dispersal patterns that are used to locate deposits using geochemical exploration techniques (Rose *et al.*, 1979). Some of the copper-bearing groundwater infiltrates to the water table, where the redox conditions change from oxidizing to reducing so the dissolved copper reacts with pyrite and other sulfides preserved by the reducing conditions to precipitate minerals that form an enrichment blanket. The enrichment blanket often contains very high copper values and is a key exploration target.

Similar energy + mass flow with feedback processes are responsible for the formation of other types of ore deposits. Ore deposits are a subset of the large number of geological and geochemical processes that are driven by energy and mass fluxes leading to pattern formation. Examples of a few that have been modeled include patterns in stylolites (Merino, 1992), agates (Wang and Merino, 1995), terra rosa (Merino and Banerjee, 2008), and burial dolomites (Merino and Canals, 2011) with many more examples given in Jamtveit and Hammer (2012), Jamtveit and Meakin (1999), and Ortoleva (1994). Creating general models that link pattern formation and unit processes like those described in this book is an exciting scientific frontier.

References

Adamson, A.W., Gast, A.P. (1997). *Physical Chemistry of Surfaces*. Wiley, New York.

Alkattan, M., Oelkers, E.H., Dandurand, J.L., Schott, J. (1998). An experimental study of calcite and limestone dissolution rates as a function of pH from −1 to 3 and temperature from 25 to 80°C. *Chemical Geology*, **151**, 199–214.

Amis, E.S. (1966). *Solvent Effects on Reaction Rates and Mechanisms*. Academic Press, New York.

Appelo, C.A.J. (1996). Multicomponent ion exchange and chromatography in natural systems. In *Reactive Transport in Porous Media*, eds. Lichtner, P.C., Steefel, C.I. Mineralogical Society of America, Washington D.C., pp. 193–227.

Applin, K.R., Lasaga, A.C. (1984). The determination of SO_4^{2-}, $NaSO_4^-$, and $MgSO_4^\circ$ tracer diffusion coefficients and their application to diagenetic flux calculations. *Geochimica et Cosmochimica Acta*, **48**, 2151–2162.

Aris, R. (1956). On the dispersion of a solute in a fluid flowing through a tube. *Proceedings of the Royal Society A*, **235**, 67–77.

Aris, R. (1989). *Elementary Chemical Reactor Analysis*. Dover Publications, New York.

Aris, R. (1994). *Mathematical Modeling Techniques*. Dover Publications, New York.

Arking, A. (1996). Absorption of solar energy in the atmosphere: Discrepancy between model and observations. *Science*, **273**, 779–782.

Arnold, F.H., Schofield, S.A., Blanch, H.W. (1986). Analytical affinity chromatography I. Local equilibrium and the measurement of association and inhibition constants. *Journal of Chromatography*, **355**, 1–12.

Arvidson, R.S., Luttgbe, A. (2010). Mineral dissolution kinetics as a function of distance from equilibrium: New experimental results. *Chemical Geology*, **269**, 79–88.

Asano, T., le Nobel, W.J. (1978). Activation and reaction volumes in solution. *Chemical Reviews*, **78**, 407–489.

Ašperger, S. (2003). *Chemical Kinetics and Inorganic Reaction Mechanisms*, 2nd edn. Kluwer Academic, New York.

Astruc, D. (1995). *Electron transfer and radical processes in transition-metal chemistry*. VCH Publishers, Inc., New York.

Augustithis, S.S., Mposkos, E., Vgenopoulos, A. (1980). Diffusion rings (sphaeroids) in bauxite. *Chemical Geology*, **30**, 351–362.

Avrahami, M., Golding, R.M. (1968). The oxidation of the sulphide ion at very low concentrations in aqueous solutions. *Journal of the Chemical Society A*, 647–651.

Avrami, M. (1939). Kinetics of phase change. I. *Journal of Chemical Physics*, **7**, 1103–1112.

Avrami, M. (1940). Kinetics of phase change. II. *Journal of Chemical Physics*, **8**, 212–224.

Axford, S.D.T. (1997). Aggregation of colloidal silica: Reaction-limited kernel, stability ratio and distribution moments. *Journal of the Chemical Society, Faraday Transactions*, **93**, 303–311.

Bahr, J.M., Rubin, J. (1987). Direct comparison of kinetic and local equilibrium formulations for solute transport affected by surface reactions. *Water Resources Research*, **23**, 438–452.

Ball, P. (1999). *The Self-Made Tapestry*. Oxford University Press, Oxford.

Bandstra, J.Z., Brantley, S.L. (2008). Surface evolution of dissolving minerals investigated with a kinetic Ising model. *Geochimica et Cosmochimica Acta*, **72**, 2587–2600.

Barenblatt, G.I. (2003). *Scaling*. Cambridge University Press, Cambridge, U.K.

Barnes, H.H. (1967). *Roughness Characteristics of Natural Channels*. U.S. Geological Survey, Washington D.C., p. 213.

Basolo, F., Pearson, R.G. (1967). *Mechanisms of Inorganic Reactions*, 2nd edn. John Wiley & Sons, New York.

Bender, E.A. (1978). *An Introduction to Mathematical Modeling*. Wiley-Interscience, New York.

Berger, G., Cadore, E., Schott, J., Dove, P.M. (1994). Dissolution rate of quartz in lead and sodium electrolyte solutions between 25 and 300°C: Effect of the nature of surface complexes and reaction affinity. *Geochimica et Cosmochimica Acta*, **58**, 541–551.

Berner, E.K., Berner, R.A. (1987). *The Global Water Cycle*. Prentice-Hall, Inc., Englewood Cliffs, N.J.

Berner, R.A. (1980). *Early Diagenesis*. Princeton University Press, Princeton.

Berner, R.A., Sjöberg, E.L., Velbel, M.A., Krom, M.D. (1980). Dissolution of pyroxenes and amphiboles during weathering. *Science*, **207**, 1205–1206.

Bethke, C.M. (2007). *Geochemical and Biogeochemical Reaction Modeling*, 2nd edn. Oxford University Press, New York.

Bevington, P.R. (1969). *Data Reduction and Error Analysis for the Physical Sciences*. McGraw-Hill, New York.

Bielski, B.H.J., Cabelli, D.E., Arudi, R.L. (1985). Reactivity of HO_2/O_2^- radicals in aqueous solution. *Journal of Chemical and Physical Reference Data*, **14**, 1041–1100.

Bishop, K.J.M., Wilmer, C.E., Soh, S., Grzybowski, B.A. (2009). Nanoscale forces and their uses in self-assembly. *Small*, **5**, 1600–1630.

Bjerklie, D.M., Dingman, S.L. (2005). Comparison of constitutive flow resistance equations based on the Manning and Chezy equations applied to natural rivers. *Water Resources Research*, **41**, W11502.

Boerlage, S.F.E., Kennedy, M.D., Bremere, I., Witkamp, G.J., Van der Hoek, J.P., Schippers, J.C. (2002). The scaling potential of barium sulphate in reverse osmosis systems. *Journal of Membrane Science*, **197**, 251–268.

Brantley, S.L., Crane, S.R., Crerar, D.A., Hellmann, R., Stallard, R. (1986a). Dissolution at dislocation etch pits in quartz. *Geochimica et Cosmochimica Acta*, **50**, 2349–2361.

Brantley, S.L., Crane, S.R., Crerar, D.A., Hellmann, R., Stallard, R. (1986b). Dislocation etch pits in quartz. In *Geochemical Processes at Mineral Surfaces*, eds. Davis, J.A., Hayes, K.F. American Chemical Society, Washington D.C., pp. 635–649.

Brezonik, P.L. (1994). *Chemical Kinetics and Process Dynamics in Aquatic Systems.* Lewis Publishers, Boca Raton F.L.

Bridgman, P.W. (1931). *Dimensional Analysis.* Yale University Press, New Haven C.T.

Bruckner, R. (2002). *Advanced Organic Chemistry Reaction Mechanisms.* Harcourt Academic Press, San Diego C.A.

Brunauer, S., Emmett, P.H., Teller, E. (1938). Adsorption of gases in multimolecular layers. *Journal of the American Chemical Society*, **60**, 309–319.

Bunge, H.J. (1997). Some remarks on modelling and simulation of physical phenomena. *Textures and Microstructures*, **28**, 151–165.

Burd, A.B., Jackson, G.A. (2009). Particle aggregation. *Annual Reviews of Marine Science*, **1**, 65–90.

Burkin, A.R. (2001). *Chemical Hydrometallurgy.* Imperial College Press, London.

Burrows, N.D., Yuwono, V.M., Penn, R.L. (2010). Quantifying the kinetics of crystal growth by oriented aggregation. *MRS Bulletin*, **35**, 133–137.

Burton, W.K., Cabrera, N. (1949). Crystal growth and surface structure: Part 1. *Discussions of the Faraday Society*, **5**, 33–39.

Burton, W.K., Cabrera, N., Frank, F.C. (1949). Role of dislocations in crystal growth. *Nature*, **163**, 398–399.

Burton, W.K., Cabrera, N., Frank, F.C. (1951). The growth of crystals and the equilibrium structure of their surfaces. *Philosophical Transactions Royal Society of London*, **243**, 299–358.

Buxton, G.V., Greenstock, C.L., Helman, W.P., Ross, A.B. (1988). Critical review of rate constants for reactions of hydrated electrons, hydrogen atoms and hydroxyl radicals (•OH/•O$^-$) in aqueous solution. *Journal of Physical and Chemical Reference Data*, **17**, 513–886.

Buxton, G.V., Mulazzani, Q.G., Ross, A.B. (1995). Critical review of rate constants for reactions of transients from metal ions and metal complexes in aqueous solution. *Journal of Chemical and Physical Reference Data*, **24**, 1055–1349.

Cabrera, N., Burton, W.K. (1949). Crystal growth and surface structure: Part 2. *Discussions of the Faraday Society*, **5**, 40–48.

Cabrera, N., Vermiyea, D.A. (1958). The growth of crystals from solution. In *Growth and Perfection of Crystals*, eds. Doremus, R.H., Roberts, B.W., Turnbull, D. John Wiley & Sons, New York, pp. 393–410.

Campanario, J.M. (1995). Automatic 'balancing' of chemical equations. *Computers in Chemistry*, **19**, 85–90.

Capellos, C., Beielski, B.H.J. (1972). *Kinetic Systems.* Wiley-Interscience, New York.

Carslaw, H.S., Jaeger, J.C. (1959). *Conduction of Heat in Solids.* Oxford University Press, New York.

Casado, J., López-Quintela, M.A., Lorenzo-Barral, F.M. (1986). The initial rate method in chemical kinetics. *Journal of Chemical Education*, **63**, 450–452.

Casey, W.H. (2001). A view of reactions at mineral surfaces from the aqueous phase. *Mineralogical Magazine*, **65**, 323–337.

Casey, W.H., Swaddle, T.W. (2003). Why small? The use of small inorganic clusters to understand mineral surface and dissolution reactions in geochemistry. *Reviews in Geophysics*, **41**, 1008.

Casey, W.H., Westrich, H.R., Banfield, J.F., Ferruzzi, G., Arnold, G.W. (1993). Leaching and reconstruction at the surface of dissolving chain-silicate minerals. *Nature*, **366**, 253–256.

Cazes, J., Scott, R.P.W. (2002). *Chromatography Theory*. Marcel Dekker, Inc., New York.

Ceccarello, S., Black, S., Read, D., Hodson, M.E. (2004). Industrial radioactive barite scale: Suppression of radium uptake by introduction of competing ions. *Minerals Engineering*, **17**, 323–330.

Chaïrat, C., Schott, J., Oelkers, E.H., Lartigue, J.-E., Harouiya, N. (2007). Kinetics and mechanism of natural fluroapatite dissolution at 25°C and pH from 3 to 12. *Geochimica et Cosmochimica Acta*, **71**, 5901–5912.

Chaudhry, M.H. (2008). *Open-Channel Flow*. Springer, New York.

Chermak, J.A., Rimstidt, J.D. (1990). Hydrothermal transformation rate of kaolinite to muscovite/illite. *Geochimica et Cosmochimica Acta*, **54**, 2979–2990.

Chotantarat, S., Ong, S.K., Sutthirat, C., Osathaphan, K. (2011). Effect of pH on transport of Pb^{2+}, Mn^{2+}, Zn^{2+}, and Ni^{2+} through lateritic soil: Column experiments and transport modeling. *Journal of Environmental Sciences*, **23**, 640–648.

Cölfen, H., Antonietti, M. (2008). *Mesocrystals and Nonclassical Crystallization*. John Wiley & Sons, Chichester, U.K.

Colombani, J. (2008). Measurement of the pure dissolution rate constant of a mineral in water. *Geochimica et Cosmochimica Acta*, **72**, 5634–5640.

Crank, J. (1975). *The Mathematics of Diffusion*. Oxford University Press, New York.

Crutchfield, J.P. (2012). Between order and chaos. *Nature Physics*, **8**, 17–24.

Cukierman, S. (2006). Et tu, Grotthuss! and other unfinished stories. *Biochimica et Biophysica Acta*, **1725**, 876–885.

Curti, E. (1997). *Coprecipitation of Radionucleides: basic concepts, literature review, and first applications*. Wettingen, Switzerland, p. 107.

Curti, E. (1999). Coprecipitation of radionuclides with calcite: Estimation of partition coefficients based on a review of laboratory investigations and geochemical data. *Applied Geochemistry*, **14**, 433–445.

Cussler, E.L. (2009). *Diffusion*, 3rd edn. Cambridge University Press, Cambridge.

Cygan, R.T., Carrigan, C.R. (1992). Time-dependent Soret transport: Applications to brine and magma. *Chemical Geology*, **95**, 201–212.

Damasceno, P.F., Engel, M., Glotzer, S.C. (2012). Predictive self-assembly of polyhedra into complex structures. *Science*, **337**, 453–457.

Danckwerts, P.V. (1953). Continuous flow systems. Distribution of residence times. *Chemical Engineering Science*, **2**, 1–13.

Davidson, L.E., Shaw, S., Benning, L.G. (2008). The kinetics and mechanism of schwertmannite transformation to goethite and hematite under alkaline conditions. *American Mineralogist*, **93**, 1326–1337.

de Groot, S.R., Mazur, P. (1984). *Non-equilibrium Thermodynamics*. Dover Publications, New York.

de Pablo, J., Casas, I., Giménez, J., Molera, M., Rovira, M., Duro, L., Bruno, J. (1999). The oxidative dissolution mechanism of uranium dioxide. I. The effect of temperature in hydrogen carbonate medium. *Geochimica et Cosmochimica Acta*, **63**, 3079–3103.

Debessy, J., Pagel, M., Beny, J.-M., Christensen, H., Hickel, B., Kosztolanyi, C., Poty, B. (1988). Radiolysis evidenced by H_2–O_2 and H_2-bearing fluid inclusions in three uranium deposits. *Geochimica et Cosmochimica Acta*, **52**, 1155–1167.

Deming, W.E. (1943). *Statistical Adjustment of Data*. Dover Publications, New York.

Denbigh, K.G., Turner, J.C.R. (1984). Residence-time distributions, mixing, and dispersion. In *Chemical Reactor Theory: An Introduction*, 3rd edn. Cambridge University Press, New York, pp. 81–110.

Denny, M.W. (1993). *Air and Water*. Princeton University Press, Princeton.

Derome, D., Cathelineau, M., Lhomme, T., Cuney, M. (2003). Fluid inclusion evidence of the differential migration of H_2 and O_2 in the McArthur River unconformity-type uranium deposit (Saskatchewan, Canada). Possible role on post-ore modifications of the host rocks. *Journal of Geochemical Exploration*, **78**–79, 525–530.

Dietzel, M. (2000). Dissolution of silicates and the stability of polysilicic acid. *Geochimica et Cosmochimica Acta*, **64**, 3275–3281.

Doerner, H.A., Hoskins, W.M. (1925). Co-precipitation of radium and barium sulfates. *Journal of the American Chemical Society*, **47**, 662–675.

Dominé, F., Dounaceur, R., Scacchi, G., Marquaire, P.-M., Dessort, D., Pradier, B., Brevart, O. (2002). Up to what temperature is petroleum stable? New insights from a 5200 free radical reactions model. *Organic Geochemistry*, **33**, 1487–1499.

Dotson, N.A., Galván, R., Laurence, R.L., Tirrel, M. (1996). *Polymerization Process Modeling*. VCH Publishers, Inc., New York.

Douglas, J.F. (1969). *Dimensional Analysis for Engineers*. Sir Isaac Pitman & Sons Ltd., London.

Dove, P.M. (1994). The dissolution kinetics of quartz in sodium chloride solutions at 25°C to 300°C. *American Journal of Science*, **294**, 665–712.

Dove, P.M., Han, N., De Yoreo, J.J. (2005). Mechanisms of classical crystal growth theory explain quartz and silicate dissolution behavior. *Proceedings of the National Academy of Sciences*, **102**, 15357–15362.

Draganic, I.G. (2005). Radiolysis of water: A look at its origin and occurrence in the nature. *Radiation Physics and Chemistry*, **72**, 181–186.

Dria, M.A., Bryant, S.L., Schechter, R.S., Lake, L.W. (1987). Interacting precipitation/dissolution waves: The movement of inorganic contaminants in groundwater. *Water Resources Research*, **23**, 2076–2090.

Drljaca, A., Hubbard, C.D., van Dldik, R., Asano, T., Basilevsky, M.V., le Nobel, W.J. (1988). Activation and reaction volumes in solution. 3. *Chemical Reviews*, **98**, 2167–2298.

Druschel, G.K., Hamers, R.J., Luthe IIIr, G.W., Banfield, J.F. (2003). Kinetics and mechanism of trithionate and tetrathionate oxidation at low pH by hydroxyl radicals. *Aquatic Geochemistry*, **9**, 145–164.

Dunning, W.J. (1955). Theory of crystal nucleation from vapor, liquid, and solid systems. In *Chemistry of the Solid State*, ed. Garner, W.E. Academic Press, Inc., New York, pp. 159–183.

Dunning, W.J. (1969). General and theoretical introduction. In *Nucleation*, ed. Zettlemoyer, A.C. Marcel Dekker, Inc., New York, pp. 1–67.

Dutrizac, J.E. (2002). Calcium sulphate solubilities in simulated zinc processing solutions. *Hydrometallurgy*, **2002**, 109–135.

Eckert, C.A. (1972). High pressure kinetics in solution. *Annual Review of Physical Chemistry*, 239–264.

Edwards, J.O., Green, E.F., Ross, J. (1968). From stoichiometry and rate law to mechanism. *Journal of Chemical Education*, **45**, 381–385.

Edzwald, J.K., Upchurch, J.B., O'Melia, C.R. (1974). Coagulation in estuaries. *Environmental Science and Technology*, **8**, 58–63.

El-Kadi, A., Plummer, L.N., Aggarwal, P. (2011). NETPATH-WIN: An interactive user version of the mass-balance model, NETPATH. *Groundwater*, **49**, 593–599.

Epstein, I.R., Kustin, K., DeKepper, P., Orbán, M. (1983). Oscillating chemical reactions. *Scientific American*, **248**, 112–123.

Erdemir, D., Lee, A.Y., Myerson, A.S. (2009). Nucleation of crystals from solution: Classical and two-step models. *Accounts of Chemical Research*, **42**, 621–629.

Erdey-Grúz, T. (1974). *Transport Phenomena in Aqueous Solutions*. John Wiley & Sons, New York.

Ershov, B.G., Gordeev, A.V. (2008). A model for radiolysis of water and aqueous solutions. *Radiation Physics and Chemistry*, **77**, 928–935.

Evans, M.G., Polanyi, M. (1935). Some applications of the transition state method to the calculation of reaction velocities, especially in solution. *Transactions of the Faraday Society*, **31** 875–894.

Eyring, H. (1935). The activated complex in chemical reactions. *Journal of Chemical Physics*, **3**, 107–115.

Feth, J.H., Robertson, C.E., Polzer, W.L. (1964). *Sources of mineral constituents in water from granitic rocks, Sierra Nevada, California and Nevada*. U.S. Geological Survey.

Field, R.J., Schneider, F.W. (1989). Oscillating chemical reactions and nonlinear dynamics. *Journal of Chemical Education*, **66**, 195–204.

Fokin, V.M., Zanotto, E.D., Schmelzer, J.W.P. (2010). On the thermodynamic driving force for interpretation of nucleation experiments. *Journal of Non-Crystalline Solids*, **356**, 2185–2191.

Ford, I.J. (2004). Statistical mechanics of nucleation: A review. *Journal of Mechanical Engineering Science*, **218**, 883–899.

Fournier, R.O. (1977). Chemical geothermometers and mixing models for geothermal systems. *Geothermics*, **5**, 41–50.

Fu, L., Milliken, K.L., Sharp Jr., J.M. (1994). Porosity and permeability variations in fractured and Liesegang-banded Breathitt sandstones (Middle Pennsylvanian), eastern Kentucky: Diagenetic controls and implications for modeling dual porosity systems. *Journal of Hydrology*, **154**, 351–381.

Fubini, B. (1998). Surface chemistry and quartz hazard. *Annals of Occupational Hygiene*, **42**, 521–530.

Fueno, T. (1999). *The Transition State*. Gordon & Breach, Amsterdam, p. 329.

Furukawa, Y., Shimada, W. (1993). Three-dimensional pattern formation during growth of ice dendrites – its relation to universal law of dendritic growth. *Journal of Crystal Growth*, **128**, 234–239.

Garg, L.C., Maren, T.H. (1972). The rates of hydration of carbon dioxide and dehydration of carbonic acid at 37°C. *Biochimica et Biophysical Acta*, **261**, 70–76.

Garn, P.D. (1975). An examination of the kinetic compensation effect. *Journal of Thermal Analysis*, **7**, 475–478.

Garrels, R.M., MacKenzie, F.T. (1967). Origin of the chemical composition of some springs and lakes. In *Equilibrium Concept in Natural Water Systems*, ed. Gould, R.F. American Chemical Society, Washington D.C., pp. 222–242.

Garten, V.A., Head, R.B. (1970). Homogeneous nucleation in aqueous solution. *Journal of Crystal Growth*, **6**, 349–351.

Gauch Jr, H.G. (1993). Prediction, parsimony and noise. *American Scientist*, **81**, 468–478.

Gebauer, D., Völkel, A., Cölfen, H. (2008). Stable prenucleation calcium carbonate clusters. *Science*, **322**, 1819–1822.

Gelhar, L.W., Welty, C., Rehfeldt, K.R. (1992). A critical review of data on field-scale dispersion in aquifers. *Water Resources Research*, **28**, 1955–1974.

Gibbs, G.V., Cox, D.F., Ross, N.L., Crawford, T.D., Burt, J.B., Rosso, K.M. (2005). A mapping of the electron localization function for earth materials. *Physics and Chemistry of Minerals*, **32**, 208–221.

Gillespie, R.J., Robinson, E.A. (2006). Gilbert N. Lewis and the chemical bond: The electron pair and the octet rule from 1916 to the present day. *Journal of Computational Chemistry*, **28**, 87–97.

Gimarc, B.M. (1974). Applications of qualitative molecular orbital theory. *Accounts of Chemical Research*, **7**, 384–392.

Goldich, S.S. (1938). A study in rock weathering. *Geology*, **46**, 17–58.

Goldschmidt, V.M. (1958). *Geochemistry*. Clarendon Press, Oxford.

Gordon, L., Rowley, K. (1957). Coprecipitation of radium with barium sulfate. *Analytical Chemistry*, **29**, 34–37.

Greenberg, A.E., Clesceri, L.S., Eaton, A.D. (1992). *Standard Methods for the Examination of Water and Wastewater*, 18th edn. American Public Health Association, Washington D.C.

Grossman, R.B. (1999). *The Art of Writing Reasonable Reaction Mechanisms*. Springer, New York.

Gunnarson, I., Arnórsson, S. (2003). Silica scaling: The main obstacle in efficient use of high-temperature geothermal fluids. *International Geothermal Conference*, Reykjavík, Iceland, pp. 30–36.

Gupta, A., Yan, D.S. (2006). *Mineral Processing Design and Operation*. Elsevier, Amsterdam.

Gutmann, V. (1978). *The Donor–Acceptor Approach to Molecular Interactions*. Plenum Press, New York.

Guy, B. (1993). Mathematical revision of Korzhinskii's theory of infiltration metasomatic zoning. *European Journal of Mineralogy*, **5**, 317–339.

Hall, C., Day, J. (1977). Systems and Models: terms and basic principles. In *Ecosystem Modeling in Theory and Practice*, eds Hall, C., Day, J. Wiley-Interscience, New York, pp. 6–36.

Hall, C., Day, J., Odum, H. (1977). A circuit language for energy and matter. In *Ecosystem Modeling in Theory and Practice*, eds Hall, C., Day, J. Wiley-Interscience, New York, pp. 37–49.

Harris, R.L. (1999). *Information Graphics*. Oxford University Press, Oxford.

Harte, J. (1988). *Consider a Spherical Cow*. University Science Books, Mill Valley, C.A.

Hartman, P., Perdok, W.G. (1955a). On the relations between structure and morphology of crystals. I. *Acta Crystallographica*, **8**, 49–52.

Hartman, P., Perdok, W.G. (1955b). On the relations between structure and morphology of crystals. II. *Acta Crystallographica*, **8**, 521–524.

Hartman, P., Perdok, W.G. (1955c). On the relations between structure and morphology of crystals. III. *Acta Crystallographica*, **8**, 525–529.

Harvey, M.C., Schreiber, M.E., Rimstidt, J.D., Griffith, M.A. (2006). Scorodite dissolution kinetics: Implications for arsenic release. *Environmental Science and Technology*, **40**, 6709–6714.

He, S., Oddo, J.E., Tomson, M.B. (1994). The inhibition of gypsum and barite nucleation in NaCl brines at temperatures from 25 to 90°C. *Applied Geochemistry*, **9**, 561–567.

Heaney, P.J. (1993). A proposed mechanism for the growth of chalcedony. *Contributions to Mineralogy and Petrology*, **115**, 66–74.

Heizmann, J.J., Bessieres, J., Bessieres, A. (1986). Advances in kinetic models. *Journal de chimie physique*, **83**, 725–732.

Helfferich, F.G. (1989). The theory of precipitation/dissolution waves. *AIChE Journal*, **35**, 75–87.

Helgeson, H.C. (1968). Evaluation of irreversible reactions in geochemical processes involving minerals and aqueous solutions – I. Thermodynamic relations. *Geochimica et Cosmochimica Acta*, **32**, 853–877.

Helgeson, H.C. (1979). Mass transfer among minerals and hydrothermal solution. In *Geochemistry of Hydrothermal Ore Deposits*, second edn., ed. Barnes, H.L. John Wiley & Sons, New York, pp. 568–610.

Helgeson, H.C., Garrels, R.M., MacKenzie, F.T. (1969). Evaluation of irreversible reactions in geochemical processes involving minerals and aqueous solutions – II. Applications. *Geochimica et Cosmochimica Acta*, **33**, 455–481.

Hellmann, R., Eggleston, C.M., Hochella Jr., M.F., Crerar, D.A. (1990). The formation of leached layers on albite surfaces during dissolution under hydrothermal conditions. *Geochimica et Cosmochimica Acta*, **54**, 1267–1281.

Hem, J.D. (1985). *Study and Interpretation of the Chemical Characteristics of Natural Water*. U.S. Geological Survey, Washington D.C.

Hendricks, D. (2006). *Water Treatment Unit Processes*. Taylor & Francis/CRC Press, Boca Raton, F.L.

Henisch, H.K. (1988). *Crystals in Gels and Liesegang Rings*. Cambridge University Press, Cambridge.

Higgins, G.H. (1959). Evaluation of the ground-water contamination hazard from underground nuclear explosions. *Journal of Geophysical Research*, **64**, 1509–1519.

Hill Jr., C.G. (1977). *An Introduction to Chemical Engineering Kinetics and Reactor Design*. John Wiley & Sons, New York.

Hofmann, A. (1972). Chromatographic theory of infiltration metasomatism and its application to feldspars. *American Journal of Science*, **272**, 69–90.

Horne, R.L., Rodriguez, F. (1983). Dispersion in tracer flow in fractured geothermal systems. *Geophysical Research Letters*, **10**, 289–292.

Houston, P.L. (2001). *Chemical Kinetics and Reaction Dynamics*. Dover Publications, New York.

Huminicki, D.M.C., Rimstidt, J.D. (2009). Iron oxyhydroxide coating of pyrite for acid mine drainage control. *Applied Geochemistry*, **24**, 1626–1634.

Huntley, H.E. (1967). *Dimensional Analysis*. Dover Publications, New York.

Huss Jr., A., Lim, P.K., Eckert, C.A. (1982). Oxidation of aqueous sulfur dioxide. 2. High-pressure studies and proposed reaction mechanisms. *Journal of Physical Chemistry*, **86**, 4229–4233.

Icopini, G.A., Brantley, S.L., Heaney, P.J. (2005). Kinetics of silica oligomerization and nanocolloid formation as a function of pH and ionic strength at 25°C. *Geochimica et Cosmochimica Acta*, **69**, 293–303.

Ingold, C.K. (1969). *Structure and Mechanism in Organic Chemistry*. Cornell University Press, Ithaca.

Jamtveit, B., Hammer, Ø. (2012). Sculpting of rocks by reactive fluids. *Geochemical Perspectives*, **1**, 341–480.

Jamtveit, B., Meakin, P. (1999). *Growth, Dissolution and Pattern Formation in Geosystems*. Kluwer Academic Publishers, Dordrecht.

Jensen, W.B. (1980). *The Lewis Acid-Base Concepts*. John Wiley & Sons, New York.

Jin, L., Auerbach, S.M., Monson, P.A. (2011). Modeling three-dimensional network formation with an atomic lattice model: Application to silicic acid polymerization. *Journal of Chemical Physics*, **134**, 134703.

John, D.A. (2010). *Porphyry Copper Deposit Model*. U.S. Geological Survey, Washington D.C., p. 169.

Kashchiev, D., van Rosmalen, G.M. (2003). Review: Nucleation in solution revisited. *Crystal Research and Technology*, **38**, 555–574.

Kirby, C.S., Cravotta III, C.A. (2005a). Net alkalinity and net acidity 1: Theoretical considerations. *Applied Geochemistry*, **20**, 1920–1940.

Kirby, C.S., Cravotta III, C.A. (2005b). Net alkalinity and net acidity 2: Practical considerations. *Applied Geochemistry*, **20**, 1941–1964.

Kline, S.J. (1965). *Similitude and Approximation Theory*. McGraw-Hill Book Company, New York.

Knapp, R.B. (1989). Spatial and temporal scales of local equilibrium in dynamic fluid-rock systems. *Geochimica et Cosmochimica Acta*, **53**, 1955–1964.

Kondepudi, D., Prigogine, I. (1998). *Modern Thermodynamics*. John Wiley & Sons, Chichester, U.K.

Kondo, S., Miura, T. (2010). Reaction–diffusion model as a framework for understanding biological pattern formation. *Science*, **329**, 1616–1620.

Korzhinskii, D.S. (1970). *Theory of Metasomatic Zoning*. Clarendon Press, Oxford, London.

Krishnamurthy, E.V. (1978). Generalized matrix inverse approach for automatic balancing of chemical equations. *International Journal of Mathematical Education in Science and Technology*, **9**, 323–328.

Krug, H.-J., Brandstädter, H., Jacob, K.H. (1996). Morphological instabilities in pattern formation by precipitation and crystallization processes. *Geologische rundschau*, **85**, 19–28.

Laidler, K.J. (1987a). *Theories of Reaction Rates*. Harper & Row, New York, pp. 80–99.

Laidler, K.J. (1987b). *Chemical Kinetics*, 3rd edn. Harper & Row, New York.

Lamberto, D.J., Alverez, M.M., Muzzio, F.J. (1999). Experimental and computational investigation of the laminar flow structure in a stirred tank. *Chemical Engineering Science*, **54**, 919–942.

Larson, M.A., Garside, J. (1986). Solute clustering and interfacial tension. *Journal of Crystal Growth*, **76**, 88–92.

Lasaga, A.C. (1980a). Dynamic treatment of geochemical cycles: global kinetics. In: *Kinetics of Geochemical Processes*, eds Lasaga, A.C., Kirkpatrick, R.J. Mineralogical Society of America, Washington D.C., pp. 69–105.

Lasaga, A.C. (1980b). The kinetic treatment of geochemical cycles. *Geochimica et Cosmochimica Acta*, **44**, 815–828.

Lasaga, A.C. (1998a). *Geochemical Kinetics*. Princeton University Press, Princeton N.J.

Lasaga, A.C. (1998b). *Kinetic Theory in Earth Sciences*. Princeton, University Press, Princeton, N.J.

Lasaga, A.C., Berner, R.A. (1998). Fundamental aspects of quantitative models for geochemical cycles. *Chemical Geology*, **145**, 161–175.

Lasaga, A.C., Blum, A.E. (1986). Surface chemistry, etch pits and mineral-water reactions. *Geochimica et Cosmochimica Acta*, **50**, 2363–2379.

Lasaga, A.C., Gibbs, G. (1990). Ab-initio quantum mechanical calculations of water-rock interactions: Adsorbtion and hydrolysis reactions. *American Journal of Science*, **290**, 263–295.

Laub, R.J. (1985). Theory of chromatography. In *Inorganic Chromatographic Analysis*, ed. Macdonald, J.C. John Wiley & Sons, New York, pp. 13–186.

Le Caër, S. (2011). Water radiolysis: Influence of oxide surfaces on H_2 production under ionizing radiation. *Water*, **3**, 235–253.

Lefticariu, L., Pratt, L.A., LaVern, J.A., Schimmelmann, A. (2010). Anoxic pyrite oxidation by water radiolysis products – A potential source of biosustaining energy. *Earth and Planetary Science Letters*, **292**, 57–67.

Leifer, A. (1988). *The Kinetics of Envrionmental Aquatic Photochemistry*. American Chemical Society, Washington D.C.

Lerman, A. (1979). *Geochemical Processes: Water and Sediment Environments*. Wiley-Interscience, New York.

Lerman, A., Wu, L. (2008). Kinetics of global geochemical cycles. In *Kinetics of Water-Rock Interaction*, eds Brantley, S.L., Kubicki, J.D., White, A.F. Springer, New York, pp. 655–736.

Leubner, I.H. (2000). Particle nucleation and growth models. *Current Opinion in Colloid and Interface Science*, **5**, 151–159.

Levenspiel, O. (1972a). *Chemical Reaction Engineering*. John Wiley & Sons, New York.

Levenspiel, O. (1972b). *Single Ideal Reactors, Chemical Reaction Engineering*, 2nd edn. John Wiley & Sons, New York, pp. 91–117.

Levich, V.G. (1962). *Physicochemical Hydrodynamics*. Prentice-Hall, Inc., Englewood Cliffs, N.J.

Lewis, G.N. (1916). The atom and the molecule. *Journal of the American Chemical Society*, **38**, 762–785.

Li, Y.-H., Gregory, S. (1974). Diffusion of ions in sea water and in deep-sea sediments. *Geochimica et Cosmochimica Acta*, **38**, 703–714.

Liesegang, R.E. (1913). *Geologische Diffusionen*. Verlag von Theodor Steinkopff, Dresden.

Liesegang, R.E. (1915). *Die Achate*. Verlag von Theodor Steinkopff, Dresden.

Limerinos, J.T. (1970). *Manning Coefficient from Measured Bed Roughness in Natural Channels*, U.S. Geological Survey Water Supply Paper 1898-B.

Lin, L.-H., Slater, G.F., Lollar, B.S., Lacrampe-Couloume, G., Onstott, T.C. (2005). The yield and isotopic composition of radiolytic H_2, a potential energy source for the deep subsurface biosphere. *Geochimica et Cosmochimica Acta*, **69**, 893–903.

Liu, J., Aruguete, D.M., Jinschek, J.R., Rimstidt, J.D., Hochella Jr., M.F. (2008). The non-oxidative dissolution of galena nanocrystals: Insights into mineral dissolution rates as a function of grain size, shape, and aggregation state. *Geochimica et Cosmochimica Acta*, **72**, 5984–5996.

Liu, L., Guo, Q.-X. (2001). Isokinetic relationship, isoequilibrium relationship, and enthalpy–entropy compensation. *Chemical Reviews*, **101**, 673–695.

Liu, S.-T., Nancollas, G.H. (1971). The kinetics of dissolution of calcium sulfate dihydrate. *Journal of Inorganic and Nuclear Chemistry*, **33**, 2311–2316.

Lowell, S., Shields, J.E. (1991). *Powder Surface Area and Porosity*, 3rd edn. Chapman & Hall, London.

MacInnis, I.N., Brantley, S.L. (1993). Development of etch pit size distributions on dissolving minerals. *Chemical Geology*, **105**, 31–49.

Maffezzoli, A., Kenny, J.M., Torre, L. (1995). On the physical dimensions of the Avrami constant. *Thermochimica Acta*, **269/270**, 185–190.

Mandel, J. (1964). *The Statistical Analysis of Experimental Data*. Dover Publications, New York.

Marangoni, A.G. (1998). On the use and misuse of the Avrami equation in characterization of the kinetics of fat crystallization. *Journal of the American Oil Chemists' Society*, **75**, 1465–1467.

Marcus, R.A. (1964). Chemical and electrochemical electron-transfer theory. *Annual Reviews of Physical Chemistry*, 155–196.

Marcus, R.A. (1968). Theoretical relations among rate constants, barriers, and Brønsted slopes of chemical reactions. *Journal of Physical Chemistry*, **72**, 891–898.

Marcus, R.A. (1985). Electron transfers in chemistry and biology. *Biochimica et Biophysica Acta*, **811**, 265–322.

Marcus, R.A. (2000). Tutorial on rate constants and reorganization energies. *Journal of Electroanalytical Chemistry*, **483**, 2–6.

Marin, G.B., Yablonsky, G.S. (2011). *Kinetics of Chemical Reactions*. Wiley-VCH, Weinheim, Germany.

Markov, I.V. (2003). *Crystal Growth for Beginners*. World Scientific, London.

McCabe, W.L., Smith, J.C., Harriott, P. (1993). *Unit Operations of Chemical Engineering*. McGraw-Hill, Inc., New York.

McClamroch, N.H. (1980). *State Models of Dynamic Systems*. Springer-Verlag, New York.

McIntire, W.L. (1963). Trace element partition coefficients: A review of theory and applications to geology. *Geochimica et Cosmochimica Acta*, **27**, 1209–1264.

Meldrum, F.C., Sear, R.P. (2008). Now you see them. *Science*, **322**, 1802–1103.

Merino, E. (1992). Self-organization in stylolites. *American Scientist*, **80**, 466–473.

Merino, E., Banerjee, A. (2008). Terra rosa genesis, implications for karst, and eolian dust: A geodynamic thread. *Journal of Geology*, **116**, 62–75.

Merino, E., Canals, A. (2011). Self-accelerating dolomite-for-calcite replacement: Self-organized dynamics of burial dolomitization and associated mineralization. *American Journal of Science*, **311**, 573–607.

Meyer, D. (1999). *Surfaces, Interfaces, and Colloids*, 2nd edn. John Wiley & Sons, New York.

Miller, D.G. (1982). *Estimation of Tracer Diffusion Coefficients of Ions in Aqueous Solution*. Lawrence Livermore Laboratory, Livermore, C.A.

Millero, F.J. (2001). The oxidation of hydrogen sulfide in natural waters. In *The Physical Chemistry of Natural Waters*. John Wiley & Sons, Inc., New York, pp. 582–632.

Moore, J.W., Pearson, R.G. (1981). *Kinetics and Mechanism*. John Wiley & Sons, New York.

Müller, S.C., Kai, S., Ross, J. (1982). Curiosities in periodic precipitation patterns. *Science*, **216**, 635–637.

Murphy, G. (1950). *Similitude in Engineering*. Ronald Press Company, New York.

Mutaftschiev, B. (2001). *The Atomistic Nature of Crystal Growth*. Springer, Berlin.

Narayanasamy, J., Kubicki, J.D. (2005). Mechanism of hydroxyl radical generation from a silica surface: Molecular orbital calculations. *Journal of Physical Chemistry B*, **109**, 21796–21807.

Nauman, E.B. (2008). Residence time theory. *Industrial & Engineering Chemistry Research*, **47**, 3752–3766.

Neta, P., Huie, R.E., Ross, A.B. (1988). Rate constants for reactions of inorganic radicals in aqueous solution. *Journal of Chemical and Physical Reference Data*, **17**, 1027–1284.

Nicholson, R.V., Gillham, R.W., Reardon, E.J. (1990). Pyrite oxidation in carbonate-buffered solution: 2. Rate control by oxide coatings. *Geochemica et Cosmochimica Acta*, **54**, 395–402.

Nico, P., Anastasio, C., Zasoski, R.J. (2002). Rapid photo-oxidation of Mn(II) mediated by humic substances. *Geochimica et Cosmochimica Acta*, **66**, 4047–4056.

Nicolis, G., Prigogine, I. (1977). *Self-Organization in Nonequilibrium Systems*. John Wiley & Sons, New York.

Nicolis, G., Prigogine, I. (1989). *Exploring Complexity*. W.H. Freeman & Company, New York.

Nielsen, A.E. (1964). *Kinetics of Precipitation*. Macmillan Company, New York.

Nkedl-Kizza, P., Rao, P.S.C., Hornsby, A.G. (1987). Influence of organic cosolvents on leaching of hydrophobic organic chemicals through soils. *Environmental Science and Technology*, **21**, 1107–1111.

Nordstrom, D.K. (1977). Thermochemical redox equilibria of Zo Bell's solution. *Geochimica et Cosmochimica Acta*, **41**, 1835–1841.

Nordstrom, D.K. (2012). Models, validation, and applied geochemistry: Issues in science, communication, and philosophy. *Applied Geochemistry*, **27**, 1899–1919.

Oelkers, E.H. (1991). Calculation of diffusion coefficients for aqueous organic species at temperatures from 0 to 350°C. *Geochimica et Cosmochimica Acta*, **55**, 3515–3529.

Oelkers, E.H., Helgeson, H.C. (1988). Calculation of the thermodynamic and transport properties of aqueous species at high pressures and temperatures: Aqueous tracer diffusion coefficients of ions to 1000°C and 5 kb. *Geochimica et Cosmochimica Acta*, **52**, 63–85.

Ohlin, C.A., Villa, E.M., Rustad, J.R., Casey, W.H. (2010). Dissolution of insulating oxide materials at the molecular scale. *Nature Materials*, **9**, 11–19.

Oreskes, N., Shrader-Frechette, K., Belitz, K. (1994). Verification, validation, and confirmation of numerical models in the Earth Sciences. *Science*, **263**, 641–646.

Ortoleva, P.J. (1994). *Geochemical Self-Organization*. Oxford University Press, New York.

Overton, W. (1977). A strategy of model construction. In *Ecosystem Modeling in Theory and Practice: An introduction with case histories*, eds, Hall, C., Day, J. Wiley-Interscience, New York, pp. 50–73.

Palandri, J.L., Kharaka, Y.K. (2004). *A compilation of rate parameters of water-mineral interaction kinetics for application to geochemical modeling*. U.S. Geological Survey, Menlo Park, C.A., p. 64.

Pauling, L. (1960). *The Nature of the Chemical Bond and the Structure of Molecules and Crystals: An Introduction to Modern Structural Chemistry*. Cornell University Press, Ithaca, N.Y. (3rd edition).

Penn, R.L. (2004). Kinetics of oriented aggregation. *Journal of Physical Chemistry B*, **108**, 12707–12712.

Penn, R.L., Tanaka, K., Erbs, J. (2007). Size dependent kinetics of oriented aggregation. *Journal of Crystal Growth*, **309**, 97–102.

Petrovich, R. (1981a). Kinetics of dissolution of mechanically comminuted rock-forming oxides and silicates – I. Deformation and dissolution of quartz under laboratory conditions. *Geochimica et Cosmochimica Acta*, **45**, 1665–1674.

Petrovich, R. (1981b). Kinetics of dissolution of mechanically comminuted rock-forming oxides and silicates – II. Deformation and dissolution of oxides and silicates in the laboratory and at the Earth's surface. *Geochimica et Cosmochimica Acta*, **45**, 1675–1686.

Pina, C.M., Putnis, A. (2002). The kinetics of nucleation of solid solutions from aqueous solutions: A new model for calculating non-equilibrium distribution coefficients. *Geochimica et Cosmochimica Acta*, **66**, 185–192.

Piscitelle, L.J. (1990). Determination of initial rates for a general class of chemical reactions: A methodology. *International Journal of Chemical Kinetics*, **22**, 683–688.

Platten, J.K., Bou-Ali, M.M., Dutrieux, J.F. (2003). Enhanced molecular separation in inclined thermogravitational columns. *Journal of Physical Chemistry B*, **107**, 11763–11767.

Plummer, L.N., Parkhurst, D.L., Thorstenson, D.C. (1983). Development of reaction models for ground-water systems. *Geochimica et Cosmochimica Acta*, **47**, 665–686.

Plummer, L.N., Prestemon, E.C., Parkhurst, D.L. (1991). *An interactive code (NETPATH) for modeling net geochemical reactions along a flow path*. U.S. Geological Survey, Reston, V.A., p. 227.

Plummer, L.N., Prestemon, E.C., Parkhurst, D.L. (1992). NETPATH: An interactive code for interpreting NET geochemical reactions from chemical and isotopic data along a flow PATH. In *7th International Symposium on Water-Rock Interaction–WRI-7*, eds Kharaka, Y.F., Maest, A.S. A.A. Balkema, Park City, U.T., pp. 239–242.

Pollard, J.H. (1977). *A Handbook of Numerical and Statistical Techniques*. Cambridge University Press, Cambridge.

Powers, J.E., Wilke, C.R. (1957). Separation of liquids by thermal diffusion. *A.I.Ch.E. Journal*, **3**, 231–222.

Presnall, D.C. (1986). An algebraic method for determining equilibrium crystallization and fusion paths in multicomponent systems. *American Mineralogist*, **71**, 1061–1070.

Prigogine, I. (1977). Time, structure and fluctuations. *Nobel Lecture December 8, 1977*, p. 23.

Prigogine, I. (1980). *From Being to Becoming*. W.H. Freeman & Company, San Francisco.

Prigogine, I., Stengers, I. (1984). *Order out of Chaos*. Bantam Books, Toronto.

Probstein, R.F. (1989). *Physicochemical Hydrodynamics: An Introduction*. Butterworths, Boston.

Purcell, E.M. (1977). Life at low Reynolds number. *American Journal of Physics*, **45**, 3–11.

Qian, J., Zhan, H., Zhao, W., Sun, F. (2005). Experimental study of turbulent unconfined groundwater flow in a single fracture. *Journal of Hydrology*, **311**, 134–142.

Rebreanu, L., Vanderborght, J.-P., Chou, L. (2008). The diffusion coefficient of dissolved silica revisited. *Marine Chemistry*, **112**, 230–233.

Rescigno, A., Thakur, A.K. (1988). Development of compartmental concepts. In *Pharmacokinetics: Mathematical and statistical approaches to metabolism and distribution of chemicals and drugs*, eds Pecile, A., Rescigno, A. Plenum Press, New York, pp. 19–26.

Rimstidt, J.D. (1997a). Gangue mineral transport and deposition. In *The Geochemistry of Hydrothermal Ore Deposits*, 3rd edn., ed. Barnes, H.L. John Wiley & Sons, New York, pp. 487–515.

Rimstidt, J.D. (1997b). Quartz solubility at low temperatures. *Geochimica et Cosmochimica Acta*, **61**, 2553–2558.

Rimstidt, J.D., Balog, A., Webb, J. (1998). Distribution of trace elements between carbonate minerals and aqueous solutions. *Geochimica et Cosmochimica Acta*, **62**, 1851–1863.

Rimstidt, J.D., Barnes, H.L. (1980). The kinetics of silica-water reactions. *Geochimica et Cosmochimica Acta*, **44**, 1683–1699.

Rimstidt, J.D., Brantley, S.L., Olsen, A.A. (2012). Systematic review of forsterite dissolution data. *Geochimica et Cosmochimica Acta*, **99**, 159–178.

Rimstidt, J.D., Newcomb, W.D. (1993). Measurement and analysis of rate data: The rate of reaction of ferric iron with pyrite. *Geochimica et Cosmochimica Acta*, **57**, 1919–1934.

Rimstidt, J.D., Vaughan, D.J. (2003). Pyrite oxidation: A state-of-the-art assessment of the reaction mechanism. *Geochimica et Cosmochimica Acta*, **67**, 873–880.

Robie, R.A., Hemmingway, B.S. (1995). *Thermodynamic Properties of Minerals and Related Substances at 298.15 K and 1 Bar (10^5 Pascals) Pressure and at Higher Temperatures*. U.S. Geological Survey, Washington D.C., p. 461.

Robinson, R.A., Stokes, R.H. (1959). *Electrolyte Solutions*, 2nd edn. Dover Publications, New York.

Rose, A.W., Hawkes, H.E., Webb, J.S. (1979). *Geochemistry in Mineral Exploration*. Academic Press, London.

Ross, J., Müller, S.C., Vidal, C. (1988). Chemical waves. *Science*, **240**, 460–465.

Rothbaum, H.P., Rohde, A.G. (1979). Kinetics of silica polymerization and deposition from dilute solutions between 5 and 180°C. *Journal of Colloid and Interface Science*, **71**, 533–559.

Rubinow, S.I. (1975). Tracers in physiological systems. In *Introduction to Mathematical Biology*. John Wiley & Sons, New York, pp. 104–155.

Schechter, R.S., Bryant, S.L., Lake, L.W. (1987). Isotherm-free chromatography: Propagation of precipitation/dissolution waves. *Chemical Engineering Communications*, **58**, 353–376.

Schepartz, B. (1980). *Dimensional Analysis in the Biomedical Sciences*. Charles C Thomas, Springfield, I.L.

Schmidt, L.D. (1998). *The Engineering of Chemical Reactions*. Oxford University Press, New York.

Schott, J., Oelkers, E.H. (1995). Dissolution and crystallization rates of silicate minerals as a function of chemical affinity. *Pure & Applied Chemistry*, **67**, 903–910.

Schott, J., Pokrovsky, O.S., Spalla, O., Devreux, F., Gloter, A., Mielczarski, J.A. (2012). Formation, growth and transformation of leached layers during silicate minerals dissolution: The example of wollastonite. *Geochimica et Cosmochimica Acta*, **98**, 259–281.

Schulze-Makuch, D. (2005). Longitudinal dispersivity data and implications for scaling behavior. *Ground Water*, **43**, 443–456.

Shen, L., Chen, Z. (2007). Critical review of the impact of tortuosity on diffusion. *Chemical Engineering Science*, **62**, 3748–3755.

Shoemaker, C. (1977). Mathematical construction of ecological models. In *Ecosystem Modeling in Theory and Practice*, eds Hall, C., Day, J. Wiley-Interscience, New York, pp. 76–113.

Silverman, J., Dodson, R.W. (1952). The exchange reaction between the two oxidation states of iron in acid solution. *Journal of Physical Chemistry*, **56**, 846–852.

Sjöberg, E.L., Rickard, D. (1983). The influence of experimental design on the rate of calcite dissolution. *Geochimica et Cosmochimica Acta*, **47**, 2281–2285.

Sleutel, M., Maes, D., Van Driessche, A. (2012). What can mesoscopic level in situ observations teach us about kinetics and thermodynamics of crystallization? In: *Kinetics and Thermodynamics of Multistep Nucleation and Self-Assembly in Nanoscale Materials*, eds Nicolis, G., Maes, D. John Wiley & Sons, Inc. Hoboken, N.J.

Smetannikov, A.F. (2011). Hydrogen generation during the radiolysis of crystallization water in carnallite and possible consequences of this process. *Geochemistry International*, **49**, 916–924.

Söhnel, O. (1982). Electrolyte crystal-aqueous solution interfacial tensions from crystallization data. *Journal of Crystal Growth*, **57**, 101–108.

Söhnel, O., Garside, J. (1988). Solute clustering and nucleation. *Journal of Crystal Growth*, **89**, 202–208.

Southworth, B.A. (1995). *Hydroxyl radical production via the photo-Fenton reaction in natural waters*. Department of Civil and Environmental Engineering. Massachusetts Institute of Technology, p. 188.

Sparks, D.L. (1989). *Kinetics of Soil Chemical Processes*. Academic Press, Inc., San Diego.

Spinks, J.W.T., Woods, R.J. (1990). *An Introduction to Radiation Chemistry*, 3rd edn. John Wiley & Sons, New York.

Staicu, C.I. (1982). *Restricted and General Dimensional Analysis*. Abacus Press, Tunbridge Wells, Kent, U.K.

Steefel, C.I. (2008). Geochemical kinetics and transport. In *Kinetics of Water-Rock Interaction*, eds Brantley, S.L., Kubicki, J.D., White, A.F. Springer, New York, pp. 545–589.

Steinmann, P., Lichtner, P.C., Shotyk, W. (1994). Reaction path approach to mineral weathering reactions. *Clays and Clay Minerals*, **42**, 197–206.

Stumm, W. (1990). *Aquatic Chemical Kinetics*. John Wiley & Sons, New York, p. 545.

Stumm, W. (1992). *Chemistry of the Solid-Water Interface*. John Wiley & Sons, New York.

Sunagawa, I. (2005). *Crystals: Growth, Morphology, and Perfections*. Cambridge University Press, Cambridge.

Taylor, G. (1953). Dispersion of soluble matter in solvent flowing slowly through a tube. *Proceedings of the Royal Society A*, **219**, 186–203.

Taylor, J.R. (1982). *An Introduction to Error Analysis*. University Science Books, Mill Valley, C.A.

Teng, H.H., Dove, P.M., DeYoreo, J.J. (2000). Kinetics of calcite growth: Surface processes and relationships to macroscopic rate laws. *Geochimica et Cosmochimica Acta*, **64**, 2255–2266.

Tester, J.W., Worley, W.G., Robinson, B.A., Grigsby, C.O., Feerer, J.L. (1994). Correlating quartz dissolution kinetics in pure water from 25 to 625°C. *Geochimica et Cosmochimica Acta*, **58**, 2407–2420.

Tuckerman, M.E., Marx, D., Parrinello, M. (2002). The nature and transport mechanism of hydrated hydroxide ions in aqueous solution. *Nature*, **417**, 925–929.

Tufte, E.R. (2001). *The Visual Display of Quantitative Information*. Graphics Press, Cheshire, CT.

Turing, A.M. (1953). The chemical basis of morphogenesis. *Philosophical Transactions of the Royal Society B*, **237**, 37–72.

Turnbull, D., Fisher, J.C. (1949). Rate of nucleation in condensed systems. *Journal of Chemical Physics*, **17**, 71–73.

Uhlmann, D.R., Chalmers, B. (1966). The energetics of nucleation. In *Nucleation Phenomena*, eds Michaels, A.S. American Chemical Society, Washington D.C., pp. 1–13.

van Boekel, M.A.J.S. (2009). *Kinetic Modeling of Reactions in Foods*. CRC Press, Boca Raton, F.L.

van Dldik, R., Asano, T., le Nobel, W.J. (1989). Activation and reaction volumes in solution. 2. *Chemical Reviews*, **89**, 549–688.

Van Herk, J., Pietersen, H.S., Schuiling, R.D. (1989). Neutralization of industrial waste acids with olivine – the dissolution of forsteritic olivine at 40–70°C. *Chemical Geology*, **76**, 341–352.

Velbel, M.A. (1989). Weathering of hornblende to ferruginous products by a dissolution–reprecipitation mechanism: Petrography and stoichiometry. *Clays and Clay Minerals*, **37**, 515–524.

Vidal, O., Murphy, W.M. (1999). Calculation of the effect of gaseous thermodiffusion and thermogravitation processes on the relative humidity surrounding a high level nuclear waste canister. *Waste Management*, **19**, 189–198.

Vogel, S. (1994). *Life in Moving Fluids*, 2nd edn. Princeton University Press, Princeton, N.J.

Wagner, W., Pruß, A. (2002). The IAPWS Formulation 1995 for the thermodynamic properties of ordinary water substance for general and scientific use. *Journal of Physical and Chemical Reference Data*, **31**, 387–535.

Wainer, H. (2005). *Graphic Discovery*. Princeton University Press, Princeton, NJ.

Waley, S.G. (1981). An easy method for the determination of initial rates. *Biochemistry Journal*, **193**, 1009–1012.

Walton, A.G. (1969). Nucleation in liquids and solutions. In *Nucleation*, ed. Zettlemoyer, A.C. Marcel Dekker, Inc., New York, pp. 225–307.

Wang, Y., Merino, E. (1995). Origin of fibrosity and banding in agates from flood basalts. *American Journal of Science*, **295**, 49–77.

Watson, J.T.R., Basu, R.S., Sengers, J.V. (1980). An improved representation equation for the dynamic viscosity of water substance. *Journal of Physical and Chemical Reference Data*, **9**, 1255–1290.

Weber Jr., W.J., DiGiano, F.A. (1996). *Process Dynamics in Environmental Systems*. John Wiley & Sons, Inc., New York.

Wechsler, J. (1988). On aesthetics in science. In *Design Science Collection*, ed. Loeb, A.L. Birkhäuser, Boston, p. 180.

Wehrli, B. (1989). Monte Carlo simulations of surface morphologies during mineral dissolution. *Journal of Colloid and Interface Science*, **132**, 230–242.

Weinstein, L., Adam, J.A. (2008). *Guesstimation*. Princeton University Press, Princeton, N.J.

Weiss, P., Driesner, T., Heinrich, C.A. (2012). Porphyry-copper ore shells form at stable pressure-temperature fronts within dynamic fluid plumes. *Science*, **338**, 1613–1616.

Weissbart, E.J., Rimstidt, J.D. (2000). Wollastonite: Incongruent dissolution and leached layer formation. *Geochimica et Cosmochimica Acta*, **64**, 4007–4016.

Wen, C.Y. (1968). Noncatalytic heterogeneous solid fluid reaction models. *Industrial and Engineering Chemistry*, **60**, 34–54.

Weng, P.F. (1995). Silica scale inhibition and colloidal silica dispersion for reverse osmosis systems. *Desalination*, **103**, 59–67.

White, A.F., Peterson, M.L. (1990). Role of reactive-surface area characterization in geochemical kinetic models. In *ACS Symposium Series 416, Chemical Modeling of Aqueous Systems II*, eds Melchior, D.C., Bassett, R.L. American Chemical Society, Los Angeles, CA, pp. 461–475.

Williamson, M.A., Rimstidt, J.D. (1993). The rate of decomposition of the ferric-thiosulfate complex in acidic aqueous solutions. *Geochimica et Cosmochimica Acta*, **57**, 3555–3561.

Williamson, M.A., Rimstidt, J.D. (1994). The kinetics and electrochemical rate-determining step of aqueous pyrite oxidation. *Geochimica et Cosmochimica Acta*, **58**, 5443–5454.

Winfree, A.T. (1972). Spiral waves of chemical activity. *Science*, **175**, 634–636.

Wolff, G.A., Gualtieri, J.G. (1962). PBC vector, critical bond energy ratio and crystal equilibrium form. *American Mineralogist*, **47**, 562–584.

Wulff, G. (1977). On the question of the rate of growth and dissolution of crystal faces. In *Crystal Form and Structure*, ed. Schneer, C.J. Dowden, Hutchinson & Ross, Inc, Stroudsburg, P.A, pp. 43–52.

Yoneawa, C., Tanaka, Y., Kamioka, H. (1996). Water-rock reactions during gamma-ray irradiation. *Applied Geochemistry*, **11**, 461–469.

Zhang, J.-W., Nancollas, G.H. (1990). Mechanisms of growth and dissolution of sparing soluble salts. In *Mineral-Water Interface Geochemistry*, eds Hochella, M.F., White, A.F. Mineralogical Society of America, Washington D.C., pp. 365–396.

Zhang, J.-Z., Millero, F.J. (1993). The products from the oxidation of H_2S in seawater. *Geochimica et Cosmochimica Acta*, **57**, 1705–1718.

Zheng, C., Bennett, G.D. (1995). *Applied Contaminant Transport Modeling*. Van Nostrand Reinhold, New York.

Zhu, C., Anderson, G. (2002). *Environmental Applications of Geochemical Modeling*. Cambridge University Press, Cambridge.

Zhu, C., Lu, P. (2009). Alkali feldspar dissolution and secondary mineral precipitation in batch systems: 3. Saturation states of product minerals and reaction paths. *Geochimica et Cosmochimica Acta*, **73**, 3171–3200.

Zlokarnik, M. (1991). *Dimensional Analysis and Scale-up in Chemical Engineering*. Springer-Verlag, New York.

Index

absorbed dose, 96
acceptor, 92
acid mine drainage, 10
acidity, 16
activated complex, 37, 79, 83, 86, 88, 89, 90, 92, 100, 101
activation
   energy, 80, 86
   volume, 90, 91
activity
   coefficient, 88
   ratio, 46
adsorbate, 107
advection, 128
affinity, 46, 123
agate, 207
aggregation rate, 193
alkalinity, 16
antilog transformation, 27
apparent rate constant, 38
Arrhenius equation, 30, 72, 84, 135, 147
Avrami equation, 197
   classical, 199
   modified, 199

balanced chemical reactions, 9
barite, 174, 191
batch reactor, 56
   continuous tracer, 63
   rate, 64, 66
   sampled, 74
   spike tracer, 60
BCF theory, *see* Burton–Cabrera–Frank (BCF) theory
Belousov–Zhabotinsky reaction, 207
BET surface area, 111
blunders, 21
Boltzmann constant, 82, 189, 193
boundary layer, 141

box model, 160
Bray–Liebhafsky reaction, 207
breakthrough, 179
Briggs–Rauscher reaction, 207
Brönsted–Bjerrum equation, 88
Brownian motion, 193
Brunauer–Emmett–Teller (BET) method, 111
Buckingham π theorem, 20, 129
buoyancy force, 129
Burton–Cabrera–Frank (BCF) theory, 126, 182

calcite, 174, 176
carbon dioxide, 15, 49
catalyst, 101
characteristic
   length, 130, 158
   time, 42, 53
Chauvenet's criterion, 22
chemical
   bond, 97
   potential, 46, 122
   reactor, 56
   weathering, 6
chord method, 65, 70
chromatography model, 178
coating
   growth model, 147
   thickness, 149
coefficient of determination, 29
compensation effect, 87
complimentary error function, 139
composite reaction, 37
computational error, 4
conductivity, 17
conservation of mass, 47, 56
continuity equation, 59, 61, 63, 156
critical nucleus, 189, 192

Damköhler number, 132, 140, 141, 156
Darcy velocity, 128, 165
Darcy's law, 129
Debye–Hückel limiting law, 88
degree of saturation, 189
dendrite, 127
denudation, 6
detailed balance, 45, 48, 50, 122
diffusion, 62, 74, 132
diffusion coefficient, 134, 147
   ion pair, 138
   mutual, 138, 139
   tracer, 134
diffusion-limited aggregation, 193
dimension, 17
dimensional analysis, 17
dimensionless numbers, 18, 20, 119, 129
discharge, 128
dispersion, 62, 74
distribution coefficient, 110, 173
Doerner–Hoskins model, 175
donor, 92
donor–acceptor theory, 98, 99
double log method, 72
Dufour effect, 206
dynamic viscosity, 128

effective diameter, 103, 104, 116
Eh, 17
Einstein, 94
electron
   localization function, 98
   transfer, 92, 96
   tunneling, 93
electrophile, 98
elementary reaction, 37, 97, 101
enrichment blanket, 208
entropy production, 206
error, 21
error propagation, 24
etch pit, 123, 126
extent of reaction, 38, 169

ferric thiosulfate, 83, 86, 89, 101
Fick's laws, 132, 142
fluorapatite, 73
flux, 37, 102, 124, 144, 148
   diffusion, 140
   reaction, 140
formal error, 4

forsterite, 105, 114, 116, 121
fraction reacted, 40, 42, 115, 116, 120, 153, 174, 177, 198
fraction remaining, 40, 42, 44, 120, 174, 177
fractional surface coverage, 109, 110
Franck–Condon principle, 79
free energy of activation, 84, 92
free radical, 96

geometric surface area, 104
Grotthuss mechanism, 134
Grotthuss–Draper law, 94
growth units, 127
guess method, 64
gypsum, 139, 141, 144, 146

half-life, 42, 44, 53, 55
hardness, 17
hillock, 123
homogeneous
   dimensions, 18
   units, 18
Hook's law, 92
hot spring, 60, 61
hydrated electron, 95
hydrogen sulfide, 43
hydrostatic head, 129
hydroxyl free radicals, 96

illite, 30
incongruent dissolution, 78
Ingold mechanism classification, 98
inhibitor, 101
initial rate method, 28, 65
International System of Units, 13
ionic strength, 88, 101
iron oxyhydroxide coating, 150

JMAK equation, 197
Johnson-Mehl-Avrami-Kolmogorov (JMAK) equation, 197

kaolinite, 30
kinematic viscosity, 128, 147
kink, 126
Kossel crystal model, 123

Langmuir
   isotherm, 112
   model, 107

leached layer, 126, 127
ledge, 126
Lewis
   acid–base theory, 98
   bond model, 97
Liesegang bands, 207
limonite pseudomorph, 153
Lineweaver–Burke equation, 109
local equilibrium assumption, 156, 167, 178
logarithm, 20
logarithmic transformation, 25, 27
longitudinal
   dispersion coefficient, 131, 159
   dispersivity, 131

Manning equation, 129
Marcus theory, 92
mass balance models, 170
mass transfer coefficient, 141, 144
   diffusion, 141
   reaction, 144
Maxwell–Boltzmann distribution, 80
mean, 21
metasomatic infiltration, 178
metastable persistence, 36
Michaelis–Menton equation, 112
microscopic reversibility, 45
mineral stability, 121
mixed flow reactor, 56
   continuous tracer, 63
   non-steady state, 75, 76
   rate, 69
   spike tracer, 61
mixing, 57, 74
molecular orbital theory, 97
molecularity, 37, 100

Nernst film model, 142
no slip condition, 141
non-equilibrium thermodynamics, 206
notation, 12
nucleation, 198
   rate, 185, 190, 191
nucleophile, 98
numerical derivative, 32, 65

oligiomer, 127
Onsager reciprocal relations, 206
opposed reaction, 47

oriented aggregation, 197
orthokinetic aggregation, 197
oscillating chemical reaction, 207
oxidation–reduction, 92

packed bed, 106
particle lifetime model, 118
partition
   coefficient, 173
   function, 80
pattern formation, 205
Péclet number, 131, 156
perikinetic aggregation, 197
periodic bond chain theory, 123
phase transfer
   coefficient, 165
   model, 164
photochemical, 93, 96
photochemistry
   first law, 94
   second law, 94
photo-oxidation, 94
photosynthesis, 94
Planck's law, 94
Plank constant, 82
plug flow reactor, 28, 56
   continuous tracer, 64
   rate, 70
   spike tracer, 62
polymerization rate, 182
polynomial method, 70
porphyry copper deposit, 208
potassium feldspar, 168
precursor complex, 92
pre-exponential, 72, 84, 87, 190
pressure, 90
pX notation, 13
pyrite, 10, 28, 33, 39, 40, 69, 70, 96, 150, 153

Q10 rule, 86
quartz, 53, 96, 121, 124, 157, 165
quasi-equilibrium model, 81
quasi-kinetic model, 156
R^2, 29

radiation chemical yield, 96
radiolysis, 95, 96
radium, 174
random error, 21

rate
    constant, 36
    equation, 36, 79
rate-determining step, 100
reaction
    intermediates, 100
    mechanism, 97, 100, 101
    order, 36, 72, 100
    progress variable, 39
    rate, 36
    reaction-limited aggregation, 194
reactive intermediates, 100
reconstruction reaction, 126
regression model, 27
reorganization energy, 93
reservoir, 160
residence time, 163
retardation factor, 179
reversible reaction, 52
Reynolds number, 130
rounding numbers, 23

salinity, 17
Schmidt number, 131, 141
schwertmannite, 204
scorodite, 66
self-assembly, 127
self-exchange reaction, 92
shrinking core model, 152
shrinking particle model, 114, 118, 198
SI units, 13
Sierra Nevada batholith, 171
significant figures, 23
silica, 53, 99, 138, 157, 183, 195
similitude, 19
site density, 105
slug, 56, 62, 74
Smoluchowski model, 193
solid–solid transformation model, 197
Soret effect, 206
specific discharge, 128
specific surface area
    BET, 111
    geometric, 104
spinodal, 186
spiral hillock, 126
stability ratio, 194
staged reactors, 56
standard deviation, 22

standard error, 22
Stark–Einstein law, 94
steady state, 160
stoichiometric coefficient, 38, 100
Stokes–Einstein equation, 136
structural error, 4
successor complex, 92
sulfur dioxide, 91
surface area/mass, 106
surface area/volume, 106
surface
    catalysis, 111
    concentration, 109
    free energy, 102, 186
    reactions, 106
    retreat rate, 114
    site, 105
surficial velocity, 128
systematic error, 21

TDS, 17
terrace, ledge, and kink (TLK) model, 126, 182
tetrathionate, 83
time constant, 43, 162
time–temperature–transformation (TTT) diagram, 200
TLK model, 126
tortuosity, 139
total dissolved solids, *see* TDS
tracer, 179
    continuous, 59
    spike, 59
tracer kinetics, 59
transition state, 37, 46, 80, 81, 90, 97, 100, 101
transition-state theory, 79
transmission coefficient, 82
TTT diagram, 200
Turing model, 207

unit analysis, 18
unit operation, 205
units, 13, 17
unopposed reaction, 39
uranium dioxide, 113

vacancy island, 126
valence bond theory, 97

validation, 5
velocity boundary layer, 141
verification, 4, 5, 9
vibrational frequency, 81
viscosity, 128
volumetric flow rate, 128

water cycle, 163

weathering, 168, 170
wollastonite, 76
Wulff
   construction, 122
   theorem, 103, 122

Zeldovich factor, 190